Praise for *The Seed Detective*

'A fascinating and original exploration of the horticultural heritage at our fingertips, and an inspiration to follow in at least some of Adam's footsteps. Informative, enlightening and entertaining but also important, *The Seed Detective* is an invitation to be inquisitive, to experiment, and to make our contribution to the plot-to-plate food movement we need to rejuvenate our relationship with food and the soil.'

MARK DIACONO, author of *Spice/a cook's companion*

'Adam's initial curiosity has taken him all over the world, collecting precious, often endangered vegetable seeds to grow and share with others.

'His passion encourages us to seek out nutritious, flavourful, local varieties instead of the lacklustre, mass-produced vegetables that have contributed to the loss of interest in this vital and delicious food source.

'One of the most inspirational books I have encountered in a long time.'

DARINA ALLEN, founder, Ballymaloe Cookery School

'If you're a vegetable growing addict or just curious about their origins, there's something for everyone in Adam's new book. Follow the Seed Detective as he searches for beans through Burmese markets, learn when a courgette is actually a marrow, and be regaled with stories about how homegrown red brussels sprouts helped decide his future. The perfect book for anyone who grows from seed!'

ROB SMITH, winner, BBC's *The Big Allotment Challenge*; garden columnist; TV presenter; seed guardian

'Reading this book feels like I am having a cup of tea with Adam in his polytunnel while he tells me fascinating stories of the plants around us. From why garlic was fed daily to the builders of the pyramids to the earliest chillies found in a Mexican cave in 6000 BCE, Adam entertains as he educates. I have a new-found respect for the humble vegetable.'

KATIE CALDESI, co-owner, Caldesi restaurants and cookery schools; best-selling author

'Whether you know your vegetables by eating them, cooking them or growing them, Adam Alexander's book will enhance those experiences with his tales. Knowing their histories, and the impact they have had on the world, gives them such a special savour that you will never think of them as mere "groceries" again.'

BARBARA DAMROSCH, author of *The Garden Primer*
and *The Four Season Farm Gardener's Cookbook*

'True seeds always have stories, and stories are what make Alexander's book so enjoyable.... At once explorer, treasure hunter, researcher and sleuth, Alexander brings a joy to his lifetime quest that is contagious.'

CR LAWN, founder, Fedco Seeds

'Written from the viewpoint of the humble seed, Alexander takes us on a journey that opens the true origin of the vegetables we put on our plates, not just addressing their heritage but also their fundamental role in the cultures of the world. They are the centrepieces of our lives. This book, narrated with personal experience combined with a deep passion, will make you think about the humble veg we consume in a completely different light and my advice to you is to open it up and jump in.'

C.M. COLLINS, head of horticulture, Garden Organic

'When you read the fascinating stories behind the origin of seeds in *The Seed Detective*, it's easy to be enthralled by the agricultural history growing in our gardens. This book takes the reader on a journey to distant lands with a passionate gardener, seed conserver and historian to learn how early cultures survived through seed growing, saving and sharing. It's an important and rare book that will appeal to anyone who grows a food garden or loves food history.'

ELLEN ECKER OGDEN, author of *The New Heirloom Garden*

'Drawing heavily on his own explorations in distant lands, Adam Alexander recounts endless stories of the origins of most of our important food crops. He is not the first to do this, yet his detailed, insightful and entertaining accounts are truly superb.'

WILL BONSALL, author of *Will Bonsall's Essential Guide
to Radical, Self-Reliant Gardening*

THE SEED DETECTIVE

*Uncovering
the secret histories of
remarkable vegetables*

Adam Alexander

Foreword by Professor Tim Lang

Chelsea Green Publishing
White River Junction, Vermont
London, UK

Commissioning Editor: Muna Reyal
Project Manager: Angela Boyle
Copy Editor: Caroline West
Proofreader: Nikki Sinclair
Indexer: Nancy A. Crompton
Designer: Melissa Jacobson
Page Layout: Abrah Griggs

Printed in the United States of America.
First printing September 2022.
10 9 8 7 6 5 4 3 2 1   22 23 24 25 26

ISBN 978-1-91529-400-5 (UK hardcover) | ISBN 978-1-91529-408-1 (US paperback)
| ISBN 978-1-91529-401-2 (ebook) | ISBN 978-1-91529-402-9 (audio book)

Library of Congress Cataloging-in-Publication Data is available upon request.

Chelsea Green Publishing
85 North Main Street, Suite 120
White River Junction, Vermont USA

Somerset House
London, UK

www.chelseagreen.com

For Julia, Jake and Jesse

Contents

Foreword   *ix*

Introduction   1

Part One: Visitors from the East   **17**
The Tale of Four Peas or Four Tall Stories   22
A Broad Bean Far from Home   41
Orange is Not the Only Colour   56
In Search of a Welsh Leek   71
Of Caulis, Krambē and Braske   85
An Aspiring Spear   106
For the Love of a Leaf   123
Thank Goodness for Garlic   139

Part Two: Arrivals from the West   **153**
More Than Just a Fruit   159
A Very Uncommon Bean   181
The Colour of Corn   198
The Tale of Two Classy Beans   216
Some Like 'em Hot   233
Not Just for Hallowe'en   252

And Finally – Seeds of Hope   267

Acknowledgements   *275*
Glossary   *278*
Notes   *283*
Index   *293*

Foreword

This book is a joy to read; a seductive call to take veg more seriously and see them for the cultural wealth they are. The literature on food is surely enriched to have this book written by a man who has 70 varieties of tomato seed in his own seed library! Someone, moreover, who sought out – like plant hunters of old – indigenous varieties of key food seeds wherever his travels for work in TV or on holidays with family took him around the world. Whereas you or I might garner photos or swims or memories, Adam Alexander has all those but also the results of his love affair with vegetables, an affair which leads him to trawl markets, gardens and people for the infrastructure of their culinary cultures.

Packed with stories distilled from decades of trips, they are lightly shared with us; one moment we are in the Middle or Far East, the next in Latin or North America. The vegetables and tales in this book span the globe, but always return to Adam's desire to try growing some of what he finds in his ample garden. To be consumed on the spot or returned to the kitchen for a meal. His rule can be summarised as: look, ask, experiment, save, cook, judge and share. The spirit of inquiry and desire to relish what vegetables give us – taste, health, variety, pleasure – is palpable. But so is his communal practice and generosity. The seeds are not owned, patented, profited from but shared in collective agri-food acculturation. He tries, saves and gives away seeds, the common gift of the quiet revolutionaries gathered worldwide under the term 'seed savers'. His tales always centre on this or that person who gave or sent him seeds. Unsurprisingly, his focus is on searching for the seeds of what local people eat and grow for themselves.

I agree with him that veg are the unsung heroes, indeed the foundation, of the good food economy everywhere. They are the glue of social and healthy lives everywhere; perhaps only Inuits and those climates where culinary plants are nigh impossible can be excused. If so important, why, we may ask, have they come to be seen simply as the practical bit of gardening? Why don't people drool over shops with cornucopias of veg in their windows as one sees people drooling over cake or clothes shops? That's an important question, central to a human understanding of how the world has become dominated by obesity and diet-related ill health. This has happened in under a century and now means humanity has turned food from the means of existence into the prime cause of premature death and also, we must acknowledge, a major driver of water and land use, of climate change and biodiversity loss. Understanding the loss of status of vegetables and the joys of their reinstatement is actually a test case for whether we humans have our priorities right.

Partly, it must be said, the significance of vegetables has declined in the Gadarene rush to fast foods and what we now know is the self-inflicted harm of over- and malconsumption. Against that bad news story, the promises and narratives of vegetable production and consumption, such as fruit and nuts, are the good news of the agri-food system. Yet the reality, so beautifully told in these pages, is that the humble veg and their seed have been industrialised, patented, corralled into tasteless cosmetic versions of goodness.

And now with Adam, they have their twenty-first-century champion. Eat more diverse and rare varieties, and it's your everyday contribution to rebuilding biodiversity in the field and garden, and not just at their edge.

How urgently this case needs champions worldwide today. Good gardening, farming and horticulture – the difference is mostly scale and skill – for the twenty-first century is surely about getting biodiversity into the field or garden and then down our throats, whereas twentieth-century agri-capitalism and farming intensification narrowed rather than protected the diversity of plants we inherited from centuries, nay millennia, of careful cultivation. And the forces of marketing and our human propensity to favour quick pleasure marginalised the role of fruit and veg further, seemingly locking them into a position of being 'worthy' when they warrant love. This will and must change.

For decades, debate and data have mounted about the importance of preserving and enhancing biodiversity and of putting plants at the heart of good cuisines and health. Yet the politics have remained worryingly weak. When the UN won agreement for the Convention on Biological Diversity in 1992, for example, the data on declining biodiversity and the need to reverse it was known. But the decline has continued – indeed, accelerated. In the pursuit of animal (mostly cattle) and grain production, primary producers – farmers, land owners and commodity traders – have locked the dominant model of food production into a narrowing range of what they grow, shaped in part by contracts and specifications increasingly set by giant supermarkets or processors. That's why phrases such as 'the food system' have entered the language of food analysis. There is a complex web of drivers which shapes how land is used.

Step forward gardeners. They, argues Adam, can maintain the diversity of seeds we inherit from the past. They value taste and local suitability more than cosmetic characteristics. They can also provide the free or uncosted labour on which good small-scale horticulture depends. What makes this book so wonderful is that its author turns on its head the dominant discourse of the food system as driven by the pursuit of efficiency and mass commodity production at the lowest price. Instead, states our author, why don't we focus on the joys and importance of eating biodiversity? It's the gardener's take on Freud's pleasure principle. You or I might not be able to grow 70 varieties of tomato, but why not have, grow and savour more than a tasteless 'globo-tomato'? And why not experiment? Veg are fun, and the kind of fun that doesn't need an Xbox, nuclear power stations or satellites monitoring your data! Just watch the pleasure young children have in planting and nurturing seeds!

As the title conveys, this is a book of formidable detective work. Adam asks questions. He tracks down. He mulls the evidence. He teases out the uncertainties. He is not deterred by false leads. He lifts lids. He is dogged in the pursuit of truth. And, yes, there is truth and value in learning about vegetables, many of which are so easy to overlook – the carrot, the leek, the brassicas (cabbage, kale, sprouts), the lettuce and 'greens' – the stuff which deracinated cultures such as ours tends to buy and waste. He wants to rescue them from the indignity of tastelessness. There are good and not so good, indeed downright poor veg. So why not grow and eat the best tasting? You may think that this is more likely to be applied to the supposedly more esoteric but now increasingly treasured veg such as asparagus, chicory, radicchio, or chilli, newly obsessed in the West. Not so. Nothing but the best for all of us is the Alexander doctrine.

And why not? We know enough about the false economics of mass-scale farming to rethink our food supplies in an era of climate change and political volatility. We also know that just-in-time mass retailing is more vulnerable than its architects expected. The logistics industries – shipping, planes, trucks, trains – is now satellite-based and already subject to increasing disruption from ransomware and cyber-insecurity. Tell that to a seed saver!

Away from the techno-world of modern hypermarket economies, this book tells of a slower narrative pace, the centuries of seed development, the spread of tastes, the dogged experimentation of generations of growers. The possible range of what could be written about veg, seeds and diversity is immense. Instead, Adam takes us on the sleuth trail with 14 vegetables, grouped by regions. The book tells stories which could easily drown in the complexity of genetics and cultural exchange. Instead, the complexity becomes intrinsic to the style of the book which combines personal enthusiasm with erudition. He tells tales, reports and encapsulates vast knowledge but all in a style which sees the joy of vegetables and their history! One could think: what is the big deal about vegetables; they are simply plants that we eat. Well, this book will disabuse you of thinking like that, if you ever did. Plants are astonishing. Culinary vegetables perhaps even more so because they represent delicate, patient interaction of humans and plant life.

This is a gift of a book in the sense that it's a communication we get too rarely about the vegetable zone. There are tens of, hundreds of thousands of, scientific papers and books on vegetables but rarely communications this accessible, this knowledgeable and this captivating. I hope you enjoy it as much as I have and do. Read. Eat. Think. Try. Vegetable gardening can be your contribution to reinvigorating (y)our diet. For Britain, which long ago exterminated its peasantry, the style and approach portrayed in this book is true to the nobler end of peasant life... the real purpose of growing food is to enhance life. If a seed is taken, patented, cloned, then it's likely to be theft. If it's taken, grown singly for direct human consumption and shared, it's culture. Indeed, it's horticulture. And we need more horticulture. This book brings the excitement and everyday back into horticulture. It encourages us to re-engage with gardening not as TV porn – watching others do it – but as the means of existence, the genuine route to self-sufficiency – growing for one's own or household's or friends' sufficiency.

In truth, when eating, we almost always are going on a journey of which we are barely aware. Yet all human history is in what we eat (and don't eat). We British have a particular debt to the flow of seeds and tastes. The UK's history of imperialism, plus armies of diehard plant hunters and Kew Gardens having been a brilliant repository/gatekeeper for centuries, means that thousands of plants have come to and left Britain. This book's argument is that not enough of them are being eaten. Indeed, the many personal stories of Adam being intrigued by this or that plant or seed don't end in a seed bank but in him trying to grow them. This is what makes the book such a treasure. He might say this is detective work but it's actually culinary horticultural work. Seek out seeds. Grow them. Watch them. Eat the produce.

TIM LANG
Professor Emeritus of Food Policy,
City University of London
April 2022

Introduction

Autumn 1988. The room was large and the monochrome chequered lino on the floor should have been replaced years ago. The white wall tiles, chipped and scratched, needed a good scrub. This was the sight that met us when my Russian interpreter and I decided to take control of the crumbling Communist Party hotel kitchen in Donetsk, an impoverished Soviet steel and coal town in eastern Ukraine. Little did I know as I walked into that abandoned kitchen – the staff had gone on strike because the only guests were us, a foreign film crew whom they didn't think should have been staying there – that this moment was to mark the start of a journey of discovery that fundamentally changed the way I looked at and came to understand the often visceral relationship we can have with what we grow and eat.

The First Encounter

With economic collapse, a result of the political breakdown of the old Soviet Union, Donetsk's supermarket shelves were bare. The farmers markets were well stocked but with prices beyond the pockets of most citizens. With a black market ready to offer me roubles for my dollars at rates way above the official one, food shopping in the draughty central market, for me at least, was a wonderful and inexpensive adventure. It was there that I met the figure who would become the most important person in my seed-hunting life and, from then on, I would always seek out this individual when scouring food markets around the world because what she had on her stall was invariably delicious, precious and unique. Wherever I was, she would come in the form of what is, I think, a character familiar to most: an 'ideal granny' type, usually of diminutive height, but always a powerful presence.

This remarkable and wonderful breed of grower is often to be found in markets worldwide, selling vegetables and fruits that she has been cultivating for years. And because she has little money, buying expensive seed from a merchant is unthinkable. She saves seeds from her crops, which may well have been grown on her small plot of land for generations. From time to time, among those prosaic vegetables are some culinary gems, and so it was in Donetsk, where I made my first discovery. A tennis-ball-sized sweet pepper with a fiery heart. Multi-lobbed and as red as a movie star's lips, this simple fruit, *Capsicum annuum*, literally changed my life.

I didn't know what to expect from this humble Ukrainian pepper when I first took it into the kitchen but, as soon as I had a nibble, I was smitten. I have been a keen vegetable grower since I was a kid and have always kept a productive vegetable garden, whether living in a city or the countryside. Until that day I had only ever sown commercial seeds. I determined to take some seed of this unique pepper home with me and see if I could grow it the following year. I was delighted with the result and have been saving and sharing seed from many abundant harvests with other gardeners ever since.*

---

* The laws that apply to both institutions and individuals regarding the collecting of seeds from other countries, and how that has informed my collecting, are discussed in the concluding chapter.

2

Thirty years ago, I never thought of vegetables as being rare or endangered, or how they were embedded in the social traditions of their native food culture – that they had their own stories to tell. Undoubtedly that same pepper is still being grown in the Ukraine's fertile black soil by a cohort of grannies, each with their own recipe. Stuffed peppers are a national favourite.

Becoming a Seed Detective

After that serendipitous encounter in Donetsk's central market, I always found an excuse to escape and scour the local markets while filming around the world. At first, I would search for chillies, beans and tomatoes. I was not so discerning then and it took me several years to distinguish commercial varieties from local ones. Within a few short years I started to think of myself as a seed detective: someone on the trail of local varieties that, first and foremost, were delicious and which I could grow in my own garden. Working in remote places, often countries in conflict and undergoing significant social upheaval and change, I realised that many vegetables that were an intrinsic part of local diet were in great danger of being lost forever. They needed to be saved from possible extinction. Gradually I started to build a library of the varieties I had come across, either in a local market or from a farmer, a gardener or a chef I met on my travels. And as the numbers grew, I became ever more curious about how these crops had found their way into the great diversity of food cultures that enrich our culinary pleasure today. Just how long had we been living with these crops and what, if any, was their place in the human story?

Why Save Seeds?

Although my initial motivation for collecting vegetables was culinary curiosity and a desire to grow things none of my neighbours had, it wasn't long before I started to understand why seed saving is so important, especially of traditional, open-pollinated and local varieties that are not commercially available – something that I explore in more detail throughout this book. What we eat today is the result of plant domestication: a process of selection and subsequent breeding that started as a necessity when Neolithic hunter-gatherers settled on the land to farm and became seed savers about 12,000 years ago. This revolutionary change in how humans sourced their food has become something that goes to the very core of our relationship with what we grow and eat. Understanding what triggered the change from hunter-gatherer to farmer still prompts vigorous debate; however, climate change, reductions in prey (and how easy this was to hunt) and population pressures were all factors. Selecting the plants that expressed the traits the farmer wanted, saving their seeds, sharing them, and then sowing them the following year became a cornerstone of the development of agriculture and is something I have been doing ever since I stumbled across that lovely pepper in a Donetsk market. This act completes a journey which is at the very core of our survival as a species: an unbroken link for me to those first growers and an endlessly repeating and magical circle of life. I believe saving seeds from one's own crops inspires us to think more deeply about the food choices we make.

There is, however, a strange paradox in the story of our journey from the small, tribal and mobile hunter-gatherer foraging for seeds to the settled Neolithic farmer saving and growing them. Hunter-gatherers had a vast range of foods from which they could forage, but once they became settled agriculturalists, this diversity of plants became smaller because they were now reliant for survival on a limited number of domesticated crops. Thus began a process of reducing the genetic diversity of edible crops that has continued ever since. Over millennia, employing careful selection, farmers reduced the number of plants they ate from hundreds, thousands even, to a handful of species that could more easily be domesticated. They employed clever strategies, such as growing several varieties of a crop together to mitigate against the effects of weather and pests on their yield, but famine became a constant companion of the first civilisations and much human ingenuity has been invested ever since in preventing or mitigating it. Creating surpluses in good years and saving seeds literally became a matter of life and death.

Where Did It All Go Wrong?

This visceral relationship with growing food has been steadily eroded, particularly in the Western world. Fast forward to today and the British, in particular, have become completely detached from the land. Sure, we can wax lyrical about Welsh lamb or Cheddar cheese, bedrocks of British food culture, and, yes, we have our own apples, Cox's Orange Pippin and the Bramley for a crumble, to name just two, but I have been constantly surprised at how little so many people know about where their vegetables come from and the stories behind their journey onto our plates. Apart from some potatoes, what vegetable varieties can we name?

Maybe it is because we were the first industrialised nation and for over 250 years have moved further away from the land, generation after generation. Very few of us have rural family ties. This sets us apart from other nations whose people are much more closely connected to the land – they have relatives who grow stuff. We don't have to look far to see this. Visit the markets of practically any European town and there you will find, in pride of place, local varieties of seasonal fruit and vegetables, many frequently associated with a particular dish. Shoppers know about them – their names, for sure, and their provenance. They are an intrinsic part of that country's national identity, along with cheeses, charcuterie, wine, beer and cider – something the British can recognise, but not so our vegetables. Most importantly, we don't know what our taste buds are missing.

Our intimate relationship with the land and what it provides continues to be diluted by urbanisation and the loss of varieties of crops, thanks to a vertically integrated system of agriculture in which the farmer receives the least reward and takes the greatest risk. This system, which started after World War II, is dominated by a handful of multinational agro-chemical businesses and the hegemony of supermarkets. Driven by a realisation that Britain could no longer rely on others to feed itself – in the 1930s less than 30 per cent of everything we ate was produced at home – and the need to become far more self-sufficient with the advent of World War II, intensive forms of agriculture became the norm. How to get more from the land? Who could argue against a greater use of chemical fertilisers and the development of high-yielding new cultivars of crops to feed the nation? Out with the old, lower yielding, locally grown traditional varieties that were well adapted to the vagaries of the British climate and supported the local food economy, including farmers saving their own seed, and in with new, high-yielding cultivars produced by big seed companies who owned their intellectual property, thus preventing farmers from saving seed themselves.

This approach to food production – chemically intensive, high-input/high-output farming with its reliance on pesticides, fungicides and herbicides to protect crops which can only be grown in this way – came at a price that we are paying today. Reduction in soil fertility and erosion (because impoverished soil is unable to retain nutrients and water) as well as a food system that depends on monoculture – growing just one variety of crop at scale – exacerbates the problems. Restoring traditional varieties from seed produced locally of all our arable crops, including vegetables, as part of a more traditional way of growing would reverse the degradation of our land speedily and cost-effectively. Regenerating our soils so they can store greater levels of carbon and microorganisms, means that sustainably grown crops can flourish, is vital for our food security, the environment and our health. A hundred years ago gardening books included advice on how to save vegetable seeds. It was considered part of gardening life, a tradition that has now nearly died out.

So, how did we get ourselves into this fine mess? The way the world produces its food changed fundamentally with the so-called Green Revolution, which started in Mexico towards the end of World War II. Agronomist Norman Borlaug (1914–2009), often called the Father of the Green Revolution, was developing new strains of wheat in the first half of the 1940s which were to transform output in Mexico. In 1944, the country imported half its wheat. Twelve years later it was self-sufficient and soon became an exporter. Similar successes were replicated in India and Pakistan. Borlaug's strains of wheat enabled yields to double and more. There is no question that countless millions of people were saved from starvation and the world has become much better able to feed itself as a result of his remarkable work, but, inevitably, the Green Revolution has extracted an environmental and economic price which we are all paying today.

Developed agricultural economies, especially that of the USA, embraced the new high-yield cultivars too. Even the U.K., which had imported half its wheat before World War II, now grows 80 per cent of what it needs – although most is used to feed cattle. In fact, Britain's ability to feed itself peaked in the 1970s when we produced about 75 per cent of what we ate. Today this number is just 60 per cent and falling. We are still dependent on the EU for 50 per cent of our fruit and vegetables, yet the outdated colonial obsession with getting everybody else to grow our food for us seems to be back in vogue among our political leaders. We can expect to import more cheap food produced with environmentally lower standards from the rest of the world. This is a dangerous policy in a world of increased competition for food when we should be trying to become more self-sufficient.

Borlaug was a visionary who understood the limitations of his work as a plant breeder. In 1970 he was awarded the Nobel Peace Prize. In his acceptance speech he acknowledged that his work had achieved temporary success, creating a breathing space that could provide sufficient food to the end of the century. Now, half a century later, with a world population that has more than doubled from 3.7 billion to 9.8 billion (as of 2020), the Green Revolution has run out of road. Countries whose food cultures before the revolution were not principally wheat-based, which include India, Pakistan and many African nations, now have populations whose diet has become dependent on modern wheat cultivars. Mexico, where maize, the world's most important source of carbohydrate, was first domesticated, now relies on millions of tons of imports to feed itself. And these imports are the result of unsustainable, intensive agricultural techniques which are further impoverishing our soil and impacting our climate. Those societies whose food culture was not wheat-based are losing or have lost the diversity of crops that offered greater resilience and food security, further adding to the dangers of reliance on non-native foods. They are also losing their own food cultures which were built around growing native crops.

It's Not All Doom and Gloom

Far from it. Food culture shows amazing resilience, as I have witnessed in India where *desi* – local varieties – of all sorts of crops are fundamental to their cuisine. Even when certain varieties appear lost, they can still be found, as I discovered. I have been inspired by the work of organic farmers in the USA who, in recent decades, have brought back into American food culture endless vegetable varieties with their roots in both native America cuisine and 400 years of European introductions. Visiting a farmers market in any US town these days is a feast for all the senses and with that a wonderful celebration of the richness of the nation's food heritage.

The call now is for a 'greener' revolution, or as one of India's Green Revolution pioneers Mankombu Swaminathan suggests, an 'evergreen revolution'. This is a genuine green revolution that puts responsible land management, sustainability and the conservation of biodiversity front and centre of managing our food security. Farmers may be a conservative bunch, but all they have done in the last 80 years is what governments have asked them to. Now the challenge is to offer them guidance and support to re-set their ways of thinking and working. Growers love their land and care about it, but education and understanding, along with practical and economically sustainable solutions, are urgently needed.

My Seed-Detective Mission

Crammed into two fridges in the garage behind my study are jars and boxes filled with envelopes containing – at the time of writing – 499 varieties of vegetable seeds, sadly most no longer commercially available: beans and peas, tomatoes and chillies, lettuces and leeks, cabbages and radish, carrots, beetroot, parsnip, turnip, sweetcorn, onions and spinach, herbs, courgettes and squash. I like to grow at least 70 different varieties each year: firstly, because I just love eating them and, secondly, to refresh and replenish my seed stock. Some crops I grow as a seed guardian for the Heritage Seed Library – seed that will be shared among the members; others simply to share with enthusiastic and curious gardeners. My mission was and remains to save seed so that I can do this and, most importantly, to return seed to those who first shared them with me and to support the work of the Heritage Seed Library in the U.K. and other libraries and gene banks around the world.

From high summer until early winter my days are spent harvesting the dried pods of beans and peas, scooping seeds from ripe tomatoes and rotting cucumbers, then washing and drying them on every available windowsill; winnowing lettuce seeds to separate them from their cotton-wool shrouds; threshing bags of cabbage and radish seed pods by jumping up and down on them – an activity that can drive me to drink! Winter is the time to enjoy squash, spooning their seeds out of the fruit's spongy centre to be sown the following spring.

Introduction

This cycle of seeking out crops to sow, saving their seed and sharing them has come to be part of how I define myself. Like those first farmers, I keep a store of dried beans, peas and chillies; bottles of tomatoes, condiments and pickled vegetables of all types that would keep me well provisioned in the event that next year's harvest is a disaster. And with a goodly supply of home-saved seeds I will always have plenty of wonderful vegetables to grow, too.

This is important at an individual level because, as more of us save seed, so the resultant crops become adapted to local conditions. The greater genetic diversity of traditional and open-pollinated varieties compared to modern cultivars makes them better able to flourish in more diverse environments. This ability to adapt increases their resilience and means they can also have a future within a local food economy. If grown by market gardeners and small-scale horticulture, local varieties strengthen our cultural attachment to the vegetable, and it can be sold at a premium. People love to buy vegetables whose stories are as local as they are. Another benefit of saving one's own seeds is that we find ourselves the following year with seeds that germinate better and faster, resulting in plants with greater vigour and, over time, greater resilience to local weather conditions. There are also commercial opportunities. Demand for organic seeds, especially of traditional varieties, is outstripping supply globally. Therein lies an opportunity for growers to diversify, saving their own seed and selling to others. Sometimes saving one's own seed leads to the accidental or deliberate crossing of two different varieties of a species. This results in a vegetable that becomes part of our national culture, as we shall see in the story of the very British relationship with the runner bean *Phaseolus coccineus*.

Championing Heritage and Heirloom

What do I mean by heritage and heirloom? We associate heritage with people and places. In the same way, heritage seeds are connected to regions and cuisines. An heirloom is something that is usually passed down through families, generation after generation. It's the same with heirloom vegetables, which are connected to individuals and families. In the USA, the two definitions are interchangeable. Regardless of how they are described, all these types of vegetable are open-pollinated, which means they are the result of a natural process of pollination, either by insects or wind, or because of self-pollination. In the U.K., open-pollinated commercial varieties that are no longer sold or under cultivation are also classified as heritage. F1 hybrid seed (the result of controlled breeding from different parents) – if treated like open-pollinated varieties – will produce offspring that are different to the parent, which is why they are not saved.

Somewhere deep within my – if not everyone's – emotional core lurks that first farmer. I freely admit that every morning when I go into my garden, I say a cheery good morning to all the plants. I talk to them individually too, concerned that they might be a bit under the weather; praising them if they are growing well, especially at harvest time. I empathise with my veggies much as a shepherd might empathise with their sheep. I care about them. I love them. I firmly believe – though without any evidence – that this emotional bond was experienced by those same first farmers. There is not a day in the year that I cannot find something tasty and nutritious in my vegetable garden grown from home-saved seed. And with their harvesting come memories of the people and places where I first found them.

A Human Connection

I have written this book out of a desire to share my enthusiasm and love for growing and eating rare, unusual, delicious vegetables, and saving and sharing their seeds. It is through conversations with fellow gardeners and food lovers that I have come to understand how keen people are to learn more about the history of the crops on their plate; especially those that have a local and entertaining story to tell.

I have lost count of the number of times people have talked to me with pride about the pleasures they have gained from growing vegetables from their home-saved seed. Their delight in telling me of the triumphs and disasters, and above all, their pleasure in completing the circle of cultivation – sowing home-saved seeds, harvesting a crop and having fun with them in the kitchen – makes it all worthwhile. I want to see a continued resurgence in the diversity of varieties we grow and enjoy. With a more intimate and personal relationship with these Cinderellas of our food culture will emerge a greater desire to nurture our crops, eat better and enjoy more. But it is flavour that counts above all else. Locally grown, just harvested and rapidly consumed, there is not a vegetable I grow that isn't superior to anything found in a supermarket aisle.

And with this curiosity comes an enthusiasm to enjoy the delights of vegetables that are new to our taste buds. I would like to believe that you, dear reader, might come away after reading this book with a new or refreshed curiosity about where the crops that are part of our every day started life and how they became so important to our sense of self. With a better understanding and awareness of the fabulous journeys these vegetables have made from wild parent to cultivated offspring, perhaps we won't look at that plate of peas in quite the same way again, but rather with wonder. If this book becomes the start of your journey into growing (if you are able), sourcing and eating delicious, rare, endangered, old and traditional varieties, I feel I will have done my job. Seeking out crops to sow, share and save creates an unbroken thread from seed to harvest to dish, and back to seed again. It's just such a lovely thing to do, rich with narrative. You will find in the following pages – at least I hope you will – many stories about vegetables that, I trust, will make you smile, including the all-too-human story of a pea called Daniel O'Rourke.

Where It All Began

The vegetable characters you will meet in the following chapters are members of 14 species organised into two parts: those whose origins are to the east of my garden in Wales and those who come from the west. Most of the vegetables you will read about in Part One were first domesticated along the edges of the Mediterranean and parts of the Middle East, an area known as the Fertile Crescent. The vegetables in Part Two are primarily from two neighbouring regions on the other side of the Atlantic: Mesoamerica, an area that includes Central America and the southern half of Mexico, and the northern parts of South America–Peru, Ecuador and Bolivia. These parts of the world make up just three of eight regions, known today as Centres of Diversity, which were identified by the Russian plant scientist Nikolai Vavilov (1887–1943). Vavilov created the world's largest seed bank in Leningrad (now St Petersburg) early in the twentieth century – the All-Russian Research Institute of Plant Industry. The regions he identified were inhabited by the most brilliant Neolithic plant breeders the world ever knew and are of fundamental importance in the globalisation of food. Not a day passes when we do not eat something that started its life in one of them.[1] Today, Vavilov's model has been challenged and a number of additional Centres of Diversity have been added to the eight he identified, including in Australasia and Africa.[2]

Agriculture originated in hilly or mountainous, tropical or subtropical regions that, at the time of domestication, would have been rich in natural resources. We might today associate many of these regions with drought, but 12,000 years ago they were all verdant places with ample rainfall. Farmers grew plants that provided them with carbohydrate and protein; for example, maize and beans in Mesoamerica and wheat and chickpeas in the Fertile Crescent.[3] The 12,000-year journey to our table that many of these crops have since taken is more than a simple jaunt from field to fork. The veggies you will be reading about exist as the result of generations of farmers selecting crops from which to save seed and, in the last 200 years, systematic selective breeding. These activities involved the full spectrum of human behaviours, including outright theft, duplicity and skulduggery, as well as endless curiosity, native genius, determination and sheer bloody single-mindedness.

Today there is a global network of gene banks and libraries dedicated to the conservation and dissemination of vegetable seeds that are no longer commercially available. It is vitally important for global food security that the genetic diversity of our food crops is held in such places. Since so few old varieties are available commercially, a gardener needs to join these institutions to access seed. Now, seed swaps have become crucial events in my gardening calendar: full of surprises, kindred spirits and lots of different veggie varieties to try. Much of the seed I save is for seed libraries and displaced people, so they can grow crops from their native homelands.

....

If you want to eat really tasty vegetables, you have to grow them yourself or get them from someone who does. There is nothing I love more than wandering through the vegetable garden to savour a freshly pulled carrot or a crisp, spicy radish, my trouser leg the ideal place to brush off the soil before that first heavenly sniff and then bite. When you have picked a tomato from the vine, warmed by the summer sun, and then savoured it ripe and juicy in the mouth, gone on a feeding frenzy along a row of just-filled peas, happy to overdose on their sweetness and depth of flavour straight from the pod, or wiped from your chin the melted butter that smothered the freshly picked and steamed sweetcorn, saving their seed completes the perfect circle of cultivation.

Vegetables are beautiful: botanically and aesthetically, and because their evolution and our gastronomic relationships with them are fundamental to the story of our own evolution and our own sense of self. It is an aspect of our human story that most of us never contemplate. Harvest time is that moment when, like opening a photo album, I am reminded of the many incarnations of the granny grower I have met over the years, as well as why that particular vegetable mattered to her and how it is now a part of my own story.

PART ONE

Visitors from
the East

The historian Mary Beard argues that the globalisation of food started with the Romans 2,500 years ago.[1] They were the first society to export their food culture as part of their imperial 'brand'. Through their vegetables – cabbages and kale, cauliflower and broccoli, asparagus, lettuce, even leeks – Roman cuisine became part of national diets throughout its vast empire and beyond. And it is the stories of those vegetables for which we must thank the Romans that form Part One.

In researching this book, I spent much time in the company of some of the greatest observers of the natural world. Many names appear frequently in the recounting of our social and cultural relationship with vegetables from the Fertile Crescent: Herodotus (c.484–c.425 BCE), who was called the Father of History and is best known for his opus *Histories*, was also a geographer whose observations offer an insight into contemporary food culture; Pliny the Elder (23–79 CE), a Roman philosopher who lived during the reign of Nero and wrote extensively about agriculture and crops; and Dioscorides (c.40–c.90 CE), a Greek botanist whose work underpinned intellectual and academic thinking on the origins of plants and their medical and culinary purposes until the revolution in scientific thinking began in the sixteenth century. These are just three of the better-known chroniclers from antiquity.

The influence of Arabic thinking and innovation in horticulture on the development of early European food culture should never be underestimated. The creation of highly sophisticated and brilliantly engineered irrigation systems from the time of the Moorish conquest of Spain in the eighth century CE enabled fruits and vegetables such as saffron, apricots, artichokes, carob, sugar, aubergines, grapefruits, carrots, coriander and rice to be grown in the country's arid areas. These became and remain the basic ingredients in Spanish cuisine. Ibn al-'Awwām lived in the second half of the twelfth century, spending most of his life as a landowner in Seville, southern Spain. He was a highly skilled and knowledgeable agriculturalist who wrote *Kitāb al-filā-ḥah* (Book of Agriculture), arguably the finest Arabic treatment of the subject as well as one of the most important of all medieval works on it in any language.

One thousand years after the fall of Rome, European academics and philosophers still believed that the plant world had been completely described by Dioscorides in his seminal work *De materia medica*, which was written in the middle of the first century CE. It was to be decades after the Europeans encountered the New World before botanists such as Rembert Dodoens (1517–1585) initiated a major rethink about how science should classify and describe our food crops. He was preceded by botanists, including Otto Brunfels, Jerome Bock and, most significantly, Leonhart Fuchs (1501–1566), who had a profound influence on his thinking. Probably the best known of this cohort remains Carl Linnaeus (1707–1778), a name that resonates to this day. Born in Sweden, his work as a botanist, zoologist and taxonomist formalised today's system of naming living organisms. He is known as the Father of Taxonomy and is the most acclaimed scientist of his age whose genius continues to inform us to this day and, I have no doubt, for generations to come.

The start of the revolution in the globalisation of food coincided with two major technological innovations in the sixteenth century: the invention of the printing press and woodblock illustrating. Our vegetables and their place in our lives have been richly described and illustrated from Egyptian times, but botanic art, which emerged with the intention of enhancing scientific knowledge, was made accessible to the population at large for the first time with the invention of the printing press by Johannes Gutenberg (*c.*1400–1468) in the middle of the fifteenth century. We may not have always loved them, but we have always been intimately connected to the crops we grow. Now, we could admire them for their scientific, aesthetic and cultural value too. The vegetable expert was no longer only the farmer, the gardener, the apothecary or the physician. This role was taken over by collectors, botanists and agriculturalists. Horticulture became a science. Add to this the fact that the entire world was being carved up by European powers exploiting a sophisticated global communications and transport network and suddenly the whole world was eating each other's native foods. Yet this period, when our curiosity to understand the importance and wonder of edible crops dominated intellectual and cultural awareness, lasted little more than three centuries. Since the early nineteenth century growing stuff has become an industrial process. Now, all these amazing crops are for most of us simply *food*. How sad is that?

The Tale of Four Peas
or Four Tall Stories

I always eat peas with honey,
I've done it all my life.
It makes the peas taste funny
But keeps them on the knife.

Anon

It was New Year's Day, and I was in the town of Luang Prabang, once the capital of Laos and a World Heritage site. Perched above the Mekong River, the city is a tourist hot spot and famous for its night market, a mecca for visitors in search of souvenirs. A short distance away is another altogether different one frequented by locals. The Phosi market, the town's largest, sells just about everything a resident might need. I was searching for the granny grower figure I had first met in a market in Donetsk nearly 30 years previously to see what gems she might have brought from her garden to sell. Accompanying me was a long-suffering interpreter who, being young and welded to his mobile phone, had zero interest in seeking out vegetable seeds. But as a polite Lao, he obligingly trailed after me as I ferreted among piles of dry chillies and peered at small bags of seeds that, hanging from the metal frames of the stallholder's pitch,

22

looked like wholesome Christmas tree decorations. It wasn't long before I found her. Immaculately wrapped in an orange shawl over a sensible quilted coat and apron – it was winter after all – silver-haired, someone's granny, and no more than four feet tall, she sat behind a tiny stall surrounded by plastic bowls of her homegrown garlic and shallots. Hanging from hooks, little packages of seeds, some no bigger than a tennis ball, had peas in them. The seed was wrinkled and mottled brown and purple. I was intrigued. Then began a conversation I shall never forget.

Me (via translator): Are those pea seeds?

Someone's Granny (SG) in a very irritated tone replied as if speaking to a very small and stupid child: Of course.

Me: How tall do they grow? (This is an important question because traditional peas and old varieties are tall. Modern varieties are short.)

SG: Tall.

Me: Aha, taller than you? (Important because even at four foot a modern pea is considered short.)

SG, as if replying to a moron: Of course.

Me: Did you grow these yourself?

SG, now convinced that she was in the presence of a nutter: Of course.

Me: And did you grow them from seed you had saved?

SG, wondering, I am sure, how much more of this she had to put up with: Of course.

Me: And have you been growing them for a long time?

SG, with a deeply patronising tone: Of course.

Me, not sure if she only had one stock reply: Did your mother grow them?

SG: Of course.

Me: And what part of the crop do you eat?

SG, yes, she was in the presence of a madman: All of it.

I asked to buy a bag and, for a few pennies, just to get rid of me, she handed one over. I dared not ask her another question as even devout Buddhists are known to become violent at times. My instincts made me think this pea was a genuine heirloom from Luang Prabang. Before leaving the town, I visited a local seed shop. The only packets on sale were modern cultivars bred in Thailand and China. Not a single local vegetable variety. I was feeling quietly confident that I had found something special. I guessed that the pea's flowers might be purple because of the colour of the seed, but I would only find out the truth about this when I grew it.

Winter in Laos is like a good summer in Britain, only hotter, so I was confident my acquisition would enjoy growing in its new home. I sowed a short single row of the seeds in late spring and waited to see what would emerge. Firstly, the seed all germinated and then the vines grew – and grew and grew. I stretched pea netting either side of the row and then I augmented this with seven-foot pea sticks, which the vines eagerly ascended. And then they started to flower. Boy, did they flower. By mid-summer there was a sea of two-tone purple flowers which then gave way to a legion of pods. I remembered what the seller had told me – she ate all the pea – so I started to harvest them as mangetout: armfuls of them. They were utterly delicious: sweet, tender and full of flavour. The plants continued to flower. I was unable to keep up with the crop. The pods were green with a pink blush; the blush of a maiden, I thought, and as tasty in a stir-fry because they were straight from the vine. I harvested the maturing pods and ate them as shelled peas. Not as sweet and tender as peas bred to be eaten freshly podded, but nonetheless they made very good eating. At the end of the season, I saved a kilo or more of seed, which I shared widely with friends and neighbours. This wonderful find, which I have named Maiden's Blush because of the pod's delightful pink hue, will have been grown in Indo-China for generations and is now part of my own food heritage.

Ancient Origins

The common or garden pea, *Pisum sativum*, was domesticated from a small genus consisting of three wild species that are indigenous to the Mediterranean and the Near East; regions that lie within the geographical area of the Fertile Crescent. The earliest archaeological evidence of the domestication of the humble pea dates back over 8,500 years to Neolithic settlements across the Near East. The wild parent of all the peas we eat today was a twining (it liked to clamber up other plants) winter annual, *P. humile* (also known as *P. syriacum*), which tells us that the species was first identified in Syria. It grew as far west as Greece, across Anatolia and the southern Balkans, and east as far as Jordan, modern-day Israel, Syria and parts of the Nile delta. Wild peas are interfertile, meaning they can cross with each other to create new hybrids. The peas we eat today are derived from two distinct but very closely related species: *P. sativum* var. *arvense*, a pea grown to be harvested dry, which generally has small, round, smooth seeds, and *P. sativum* var. *sativum*, the green pea, and its kin, mangetout and snap peas.[1]

For thousands of years peas were grown and harvested as a dry food that could be stored for months, even years. High in protein, they provided an important hedge against crop failure and famine as well as a meat supplement.

Within 2,000 years of first being cultivated, peas became an early example of the globalisation of food, being common across the entire Fertile Crescent, northeast Africa, western Europe and the Indian subcontinent. At this time, it is believed that *P. syriacum* crossed with another species of wild climbing pea, *P. sativum* subsp. *elatius*, which grew in the moister regions of the eastern Mediterranean to give us a new hybrid, *P. sativum*, our modern pea. It subsequently found its way to China about 2,200 years ago and was in widespread cultivation by 600 CE, when it became a part of Japanese food culture. There is a third species, *P. arvense* var. *abyssinicum*, which was independently domesticated and grown in Ethiopia.[2] I have yet to have the pleasure of tasting this unique pea, which is known locally as *dekoko* (tiny seed) and *yagere ater* (Pea of My Country). By all accounts, eaten fresh the peas are very sweet with a distinctive flavour. The flowers are a decorative red/purple colour. Sadly, this species is highly endangered and is at risk of going completely out of cultivation. Yet it has been a core part of Ethiopian cuisine and food culture for thousands of years and its cultivation and use should be championed.

What's in a Name?

The cultivated pea was not considered of great culinary value by the Romans, but in *De re coquinaria* (The Art of Cooking) a collection of recipes compiled in the first century CE, there were 14 that included peas; the seeds were soaked and boiled before being added to a dish. Although peas were an important food in Britain for millennia, they only get their first mention in records at the time of the Norman Conquest nearly a thousand years ago. By then, they were under widespread cultivation, especially on monastery land. The first ever reference to a named variety of garden pea, the Hastings, was in a poem written in the middle of the fifteenth century by a Benedictine monk from Bury St Edmunds called John Lydgate. The same pea was also planted in the garden of the Archbishop of Rouen in 1486.

First recorded in England in the sixteenth century and still in cultivation today is the Carlin pea. It goes under a variety of other names, including the Pigeon pea – because monks fed them to pigeons – the Brown Badger and the Brown, Black and Maple pea, to name but a few.[3] It was widely grown in Lancashire and across northeast England, and eaten on Carlin Sunday the week before Palm Sunday. Herbals of the Renaissance describe several varieties of pea, including field peas, which were used to make split-pea soup, as well as garden peas like Hastings that were also eaten as a fresh shelled pea. However, for the most part, peas were grown to be dried until the seventeenth century, when new varieties became more widely enjoyed by affluent and discerning citizens and were eaten whole as mangetout or shelled for their fresh green seeds.

Battle of the Breeders

Cultivated peas are self-fertile inbreeders, which means they don't need pollen from another flower to be fertilised. So accidental mutation or cross-pollination and hybridisation is rare. This means that when these events do occur, viable new crosses can quickly multiply as new varieties – in other words, they are stable with identical offspring. As a result, the pea can flourish in a wide range of climates and habitats and comes in a great many guises.

27

The first record of the artificial cross-pollination of peas was by the British botanist and horticulturalist Thomas Andrew Knight in 1787. The varieties he developed were the progenitors of the seemingly endless number that have followed since. Pea breeding was a nineteenth-century obsession in Western Europe and the United States. Hundreds of different varieties were developed by an eclectic cohort of farmers, plant breeders and amateurs. The rich and poor of Victorian Britain loved growing, exhibiting and eating peas, as did the citizens of post-revolutionary France and the USA in particular, who were more than happy to invest in their new passion. It should come as no surprise, therefore, that breeding the humble pea was of such interest to numbers of conmen and thieves, who exploited each other and a gullible public.

Serious plant breeding over the next hundred years meant that by the end of the nineteenth century, the USDA (U.S. Department of Agriculture) had recorded a total of 408 varieties of pea being grown commercially. By 1983 the number recorded was just 25, a loss of over 90 per cent. This decline of genetic diversity is not unique to the pea. It is happening with all edible crops and poses an existential threat to human survival.

A Case of Mistaken Identity

The second of my four tall pea stories demonstrates perfectly how, when marketing a product, one should never let the truth get in the way of a good story. I am not alone in growing weird and wonderful vegetables, saving the seeds and then sharing them with other like-minded gardeners. Another such person is Liam Gaffney. An Irishman living in Scotland, he and I have been swapping vegetable seeds for many years. One day he asked me if I fancied growing a couple of Irish peas that he had received from the Irish Seed Savers Association, a wonderful seed library dedicated to preserving Ireland's food crop heritage. One of the varieties went by the name of Daniel O'Rourke – about as Irish a name as one could ask for – which had supposedly originated in County Cork in the early 1800s. The Irish Seed Savers Association claimed that the pea is an original Irish heritage variety which was held by the great Russian seed collector Nikolai Vavilov (1887–1943) who had it in his library in Petrograd (now St Petersburg) in 1921. Was this a story too good to be true? The question I asked myself was: did Vavilov visit Ireland to collect seeds from whoever had bred the pea, maybe a Mr O'Rourke, or did the said Mr O'Rourke find himself in Leningrad and help himself to a pea from the collection? You really cannot put anything past a plant breeder, believe me. Vavilov travelled widely but there is no record of him ever having visited Ireland, although that doesn't mean to say he didn't.

After further investigation I discovered a very different story. It appears that this particular pea is American. An English seedsman of the time, Geoffrey Charlwood, said that it was bred by a Mr Waite of Massachusetts, who put it up for sale in England in 1853. A savvy salesman, he named the pea after the Derby winner of 1852, Daniel O'Rourke, an English thoroughbred whose sire was Irish. The horse was born and trained in County Durham, England. So, not a lot of Irish provenance there, then, but a most astute marketing ploy in a country that was mad about growing new varieties as well as horse-racing! For many years afterwards, pea breeders suspected that Daniel O'Rourke was synonymous with another famous variety of the time, Sangster No.1.*This pea was bred by an Irishman, Joseph Sangster from Lissadell in County Sligo. It is highly likely, in my view, that Sangster in fact acquired Daniel O'Rourke and simply renamed it – a very common habit among breeders and seedsmen at the time. Waite claimed that Daniel O'Rourke was a cross between another US variety, Early Burlington, and one of the first truly American-bred peas, Landreth's Extra Early. Daniel O'Rourke was a great commercial success – both in the U.K., where it was first sold, and later in the United States. More than 30 years after its introduction, it was still hugely popular and considered one of the finest culinary peas available anywhere.

By 1868 both an English and an American version of the pea were being sold, so just what nationality is the one that came from the Irish Seed Savers Association? If Vavilov did acquire Daniel O'Rourke, it was most likely during one of his trips to the USA in the 1920s when he collected a huge number of seeds of different food crops. An 'improved' Daniel O'Rourke was grown for the World Fair held in Chicago in 1893, celebrating 400 years since Christopher Columbus landed in what is now the Bahamas. Later comparisons with the original Daniel O'Rourke pea showed little variation. The 'improved' variety remained true to the original description.† Victorian breeders tried to determine

---

* A plant variety that is considered synonymous with another is one that shares the same traits and general characteristics. In other words, it's the same variety but given a different name.

† Varieties were 'improved' by plant breeders as a means of increasing sales. Often this 'improvement' resulted in a more consistent set of traits which could mean a heavier crop and less variability. More

the differences between Daniel O'Rourke and Sangster No.1 and other similar varieties, including peas with names such as Early Princess and an English stalwart, Prince Albert. As the nineteenth century drew to a close, the son of America's pre-eminent seed merchant of the time, David Landreth Jr, claimed that there were very few points of difference between the original Daniel O'Rourke and his father's own Extra Early when grown together at his Idaho research station. The latest analysis by Danish researcher Svend Erik Nielsen points unequivocally to the fact that Daniel O'Rourke is indeed synonymous with Sangster No.1. Perhaps at last the case is solved.

What I love about Daniel O'Rourke and its story is that, apart from being really tasty, easy to grow and prolific, this pea has been claimed by a nation to which there is only a tenuous connection. It is very similar, identical even, to other varieties, so just how unique is it? It was showcased as a jewel of American plant breeding, yet Vavilov presumed it to be Irish nearly 40 years after the pea had reached its zenith as an American pea at the World Fair in Chicago. And why did it go out of production? Maybe in the cut-throat world of plant breeding Daniel O'Rourke just couldn't compete, so seedsmen simply stopped growing it. Tall peas were being superseded by shorter varieties that commercial growers could harvest more easily and which were becoming ever

---

often than not, nothing was done to materially improve the principal traits of the original variety, but this was a means for a seed supplier to claim some form of ownership. Another common sales trick was to reverse the word order in a name – particularly common among sweet pea breeders – to have something new to offer each year; for example, Pink Rose one year to Rose Pink the following.

more popular in kitchen gardens and on allotments. Yes, Daniel O'Rourke is delicious, but it is not the finest pea I have ever eaten. That particular beauty I will reveal later. However, Daniel O'Rourke is safe, and anyone can grow it simply by joining the Irish Seed Savers Association and buying the seed from them.

The Plot Thickens

While many of the hundreds of varieties under cultivation might have had common traits, one can only guess at what they tasted like. As the story of Daniel O'Rourke tells us, peas of the same variety were being given different names by growers who wanted to profit from the craze for garden peas that swept the world at the time. Exotic stories about the provenance and unique traits of these hundreds of varieties helped trick gullible growers into parting with their cash to unscrupulous sellers. A great example of this was the 'discovery' of some shrivelled peas in a hermetically sealed jar acquired by the British Museum early in the nineteenth century that had lain dormant for 3,000 years in an Egyptian pharaoh's tomb. This 'fact' became public in 1844. A great charlatan of the time, William Grimstone, who ran a very dodgy market garden called the Herbary in Highgate selling exotic cure-alls, claimed that, after other growers had failed, he was able to get one of the last three surviving peas to germinate. Within a couple of years, he had grown out enough seed from this one pea to sell to an eager public, who were willing to pay over the odds. However, it wasn't long before the Horticultural Society of London smelled a rat. Duped purchasers had complained that Grimstone's Egyptian Pea was no big deal and looked much like any other marrowfat pea (so named because they are a large type) of the time. In 1849, the Society ran a comparative trial with a pea unimaginatively named Dwarf Branching Marrowfat. The two varieties were shown to be identical. However, this shocking news failed to put Mr Grimstone out of business. There was a seemingly endless supply of 'new' types to appease the grower who was mad about peas, which he continued to supply.[4]

In 1861 the Horticultural Society of London became The Royal Horticultural Society and decided to test all the peas that had been submitted to it at the time. Out of the 235 varieties they grew, only 11 were considered worthy of merit, having been judged on their appearance, growing habit and flavour. Just one of those varieties, Champion of England, is available commercially today. It was brought back to life, so to speak, thanks to the efforts of gardeners and the Heritage Seed Library who held seed. Another variety, the English stalwart Prince Albert, was popular in America too and grown on Thomas Jefferson's Virginia plantation, Monticello. It can still sometimes be bought from the gift shop there. Jefferson claimed that Prince Albert was no different to another garden pea, Early Frame. The American food historian and writer William Woys Weaver thinks otherwise, claiming that Prince Albert is taller and, if anything, is the same as another very popular garden pea of the time, Early Charlton. Confusion was compounded by The Royal Horticultural Society who described Early Charlton as being no more than three feet tall, whereas other descriptions have it growing to five or six feet!

Despite the competition between growers around the world, it was to England they would have turned their envious gaze because the weather afforded excellent growing conditions. In the USDA 1937 *Yearbook of Agriculture*, the

American geneticist B.L. Wade conceded that 'the climate of England is especially favourable to the production of large-podded market-garden peas of high quality'.

A Pea with a Colourful Provenance

As a seed detective, unravelling often conflicting and contradictory narratives is enough to drive me to drink! Which brings me to the third tall pea. Another fellow seed saver, Dutchman Gerrit Oskam, told me about a variety of mangetout, Jaune de Madras, that he had been given from the Court of Eden gene bank based in Utrecht in the Netherlands. Gene banks regularly share plant material for research and development purposes. Jaune de Madras aroused my interest because it is a wonderful example of how the power of marketing can mix fact, fiction and conjecture to create a story that touches our sense of identity, regardless of where we come from. Gerrit told me that this particular pea had been shared some years ago among seed libraries and had been the basis for an 'improved' version by an American breeder. My detective's nose was twitching for two reasons. Firstly, I had been assured by Gerrit that Jaune de Madras has a sensational flavour and I wanted to see how true this was and, secondly, just where had this pea originated and how did it get its name?

Jaune de Madras was developed in France at the height of the pea-breeding frenzy in the middle of the nineteenth century by the famous seed company, Vilmorin-Andrieux. It was also one of the peas believed to have been grown by Gregor Mendel (1822–1884) for his research on genetics. Speculation about its provenance and how it got its name has assumed that the pea originated in India. Indeed,

Vilmorin-Andrieux claimed as much, and this seems plausible because Indians have been growing peas for millennia. There is a Madras yellow slate which is quarried in India and there is an official Madras yellow colour palette, which looks more green than yellow to me. Madras curry powder is distinctively yellow. The so-called 'improved' American version, Golden Sweet, allegedly bred around the same time, although only sold some years later, is also suggested by some to have originated in India. The plot thickens.

Intrigued by the stories, I grew Jaune de Madras and was not disappointed. I found it to be one of the finest flavoured mangetout I had ever eaten, as attested by everyone I gave it to. That was in 2015 when I also first found Golden Sweet, suddenly available commercially once more. Specialist seed companies selling heritage and heirloom varieties claimed it to be rare and unique with suggestions that it had been developed in the United States in the latter part of the nineteenth century. No mention of Jaune de Madras, I can only presume because to admit the French connection might dent national pride. I needed to find out for myself just how different this variety was from Jaune de Madras, so, the following year, I grew the two side by side. I thought that Jaune de Madras was prettier, with yellower stems, more translucent young pods and a far better flavour, although the habit of both varieties was identical, and my un-scientific cultivation and observation of the pods and flowers disclosed no obvious variation. So, the question I asked myself was: just what did the nameless American breeder actually do to 'improve' Jaune de Madras to give the world Golden Sweet? Perhaps it was as simple as renaming a rather excellent French variety and giving it a new nationality – American – in the hope of making it more commercially attractive. At least the Americans don't claim they 'discovered' or first bred this heavenly yellow mangetout. But which sounds better, Golden Sweet or Jaune de Madras? I prefer the latter. In any case, the American breeder would have been following a well-trodden path of doing very little to a variety in order to reinvent it for their own commercial gain. And if the belief in Indian roots for this pea is true, then what might it have originally been called there?

Frustratingly, any information about the origins of Jaune de Madras and its progeny is conflicting and speculative. In India, the traditional varieties that had been grown for millennia were round-seeded and used as a dry pulse, the same as lentils and chickpeas. It was only with the arrival of the European colonial powers in the seventeenth century that the pea we eat both shelled and as mangetout was widely cultivated. During the seventeenth and eighteenth centuries, the French colonial powers established trading stations and controlled swathes of southeast India along the Coromandel Coast, which includes modern-day Chennai (formerly Madras). For over a century, this huge area was constantly fought over by the British and French, who finally did a deal in 1749, leaving it to the British. There are no contemporary records of native peas of any colour growing in the region, which was known by the end of the eighteenth century as the Madras Presidency. So, perhaps geography is not the reason for Jaune de Madras having this name. After all, the French were long gone from India when it first appeared in their kitchen gardens in the nineteenth century. I like to think that a smart employee at Vilmorin-Andrieux invented the back story of the yellow mangetout to help sales. There is nothing remotely Indian about Madras curry either, apart from its name. It was invented by a British chef and appeared in London restaurants in the middle of the nineteenth century at the same time as Jaune de Madras was made available to pea aficionados in the U.K. I rest my case.

Back from the Brink

Serendipity plays a big part in seed hunting, which brings me to the fourth of my tall peas. I always have high hopes of coming across something unique and tasty when travelling in countries I have not visited before. So, it was wonderful to stumble upon a very special European pea. In 2014, I was on holiday in the region of Catalonia known as Garrotxa. It is a part of the southern Pyrenees which, like most regions of Spain, is fiercely proud of its food culture. Thanks to a conversation with my host about my quest to find local varieties of vegetables, he introduced me to a kindred spirit, Jesus Vargas. His small organic farm nestles in an idyllic fold of this region where he grows about 150 varieties of vegetables, almost all of which are native to Catalonia and northern Spain. Roaming freely though his vegetable garden I came across a jungle of peas. The vines were scrambling across the ground and it was impossible to tell just how tall they were. It was mid-May and they were heavy with long, fat pods. They looked irresistible and the large raw peas were unimpeachably delicious. I had to have some seed. Jesus, generous as most seed savers are, happily furnished me with a jam jar full. His wife's grandfather had bred the variety himself and named it after her grandmother, Avi Joan. This pea is a true heirloom. A romantic one, too.

Until I met Jesus and he gave me some of his precious harvest, he was the only person in the world growing this amazing – and now my favourite – pea. On my return home

I allowed the seedlings to grow through a forest of seven-foot-tall hazel pea sticks. They proved to be too short. Avi Joan grew another three feet, flopping over and thus making the picking easier. What is remarkable about this pea is that it remains sweet and very flavoursome, even as the pods mature. I have been able to share seed with other growers, so now Avi Joan is being grown in dozens of gardens across the U.K. Jesus's grandfather was one of a remarkable tribe of plant breeders and enthusiasts that for the last couple of hundred years or so have been enriching our diet with utterly divine crops. Unlike many who have appeared in this chapter already, he was an honest breeder. Catalan food culture, like that of its continental neighbours, is at the heart of its unique social history. I asked Jesus what would have happened to his pea if he had died before sharing it with me. He had no answer. Now, thankfully, it can be in the world for generations to come.

The Next Generation

Despite the domination of global seed companies and plant breeders producing endless new varieties of peas for mass production, there are still individual growers who continue to breed new varieties using traditional methods. Today a modern variety of mangetout with a purple blush is widely available. It's called Spring Blush and is the result of some very careful selection by a remarkable, traditional American breeder, Dr Alan Kapuler, who farms in Oregon. I asked him for the story behind his new pea and what he told me was an extraordinary journey of trial and error, in which he grew a succession of new varieties through crossing until he got to Spring Blush. His story began with an unhelpful encounter with one of the world's largest plant breeders. The Northrup King Seed Company is one of America's oldest seed companies and has a fearsome reputation. It won a landmark decision in 1970 to trademark seeds at the expense of small producers. The company refused to give Alan permission to purchase one of their varieties, Sugar Snap Vine Pea seeds, to grow organically to eat, so he took matters into his own hands.

If Northrup wouldn't sell him their seeds, Alan would breed his own. First, he needed to learn how to cross peas to create new varieties. His bible became an English translation of Gregor Mendel's papers on genetic inheritance while experimenting with peas, first published in 1900.* In it he found detailed drawings of the pea flower and how to go about making crosses. This provided the impetus for Alan to take up breeding peas and continue a 200-year-long American tradition. After eight years work, his pea, Spring Blush, which is not dissimilar to the Sugar Snap Vine Pea that Northrup refused to sell him, has become widely available. Being open-pollinated, anyone who grows this lovely mangetout can save their own seed.

What I find so inspiring about people like Alan Kapuler is that there are still backyard growers developing new varieties of pea, following in the footsteps of the English botanist Thomas Andrew Knight who created a hybrid pea in 1787 – probably the first time that artificial cross-pollination of any crop had ever taken place.[5] Today the battleground is

---

* Mendel, known as the Father of Modern Genetics, bred and cross-bred peas in order to understand the process of inheritance. His paper, 'Experiments on Plant Hybridization', published in 1865, included detailed descriptions of the technique for extracting pollen and embryos from pea flowers.

no longer about plagiarising varieties and renaming them to suit. On one side are the giant seed producers who want to patent their new varieties, which have only been made possible by the free flow of genetic information. These varieties are bred to be grown for the freezing and canning trade. They are uniform, fast cropping, and ready for harvest at the same time – and so not of much use to a gardener who wants their crop to mature over a long period. Ranged against the major seed producers are growers and breeders of open-pollinated heritage, heirloom and ex-commercial varieties who believe in plant breeding being 'open source', sharing their creations with everyone, just like Alan Kapuler. Long may they continue.

A Broad Bean
Far from Home

*...it was thought that the souls of the
dead were enclosed in them and that they
resembled the gates of hell.*

Pellegrino Artusi (1820–1911) –
*Science in the Kitchen and
the Art of Eating Well* (1891)

I was becoming irritated with the endless roadside checks. Every few miles, at the entrance and exit to the smallest villages, paramilitaries in leather jackets wielding AK47s and lads with ancient shotguns manned rudimentary barricades and politely but insistently asked to see our papers. It was the spring of 2011 in Syria and the first rumblings of dissent from an oppressed nation were being ruthlessly dealt with by a very nasty regime indeed. I had worked as a film-maker in many war zones over the years, so was not especially fazed by the presence of guns. Nonetheless, the experience was deeply unpleasant. I wanted to share the optimism of everyone I met in Syria at the time, who hoped that Assad would make reforms sufficient to mollify his critics. Sadly, we were all to be disappointed. Little did I realise at the time that my curiosity about local vegetable varieties would result in a number of encounters with one that has been part of British food culture for millennia.

In 2011 Syria had a vibrant and very important plant breeding sector. Successful seed companies were producing local varieties of a huge number of vegetables. As interested as I was in these commercial seeds, what I really wanted to do was to seek out heritage and heirloom varieties whose stories might take me back thousands of years to a time when they were first domesticated. Syria was, before Assad went to war against his own people, the breadbasket of the Middle East, at the heart of the Fertile Crescent.

A Syrian Favourite

The fava or broad bean, *Vicia faba*, is one of the oldest cultivated crops in the world and it was at the top of my wish list. My travels through Syria took me to Palmyra, a ruined Roman city in the centre of the country – a magnificent oasis surrounded by endless scrub, mountains and desert. When I visited, the city was devoid of tourists, who had been advised either not to come to the country or to leave as soon as possible. So, we had this amazing location pretty much to ourselves. Next to the ruined city was an empty restaurant serving the ubiquitous tourist buffet, which included a salad of fava beans. They were pale green, enormous – larger than my thumb – and delicious. I asked the chef if he had any dried beans I could try and grow back home. He gladly gave me a good handful which, you've guessed it, now grow happily in my own garden. He had been cultivating this bean for years on his farm in the oasis, as had his father and his grandfather before him.

Syria is where many of the first eight crops to be domesticated came from. Known as Founder Crops, these are einkorn and emmer wheat, barley, peas, lentils, chickpeas, flax and bitter vetch, which is closely related to the fava bean.[1] This process of domestication started about 12,000 years ago and one of the earliest archaeological sites revealing cultivation is at Tell Abu Hureyra, a settlement on the south side of the river Euphrates, just 75 miles east of Aleppo. The settlement dates back nearly 13,000 years and beans excavated there were found to be almost 10,000 years old. Sadly, this remarkable place has been inundated by the waters of Lake Assad after the Tabqa Dam was completed in 1974.

To date, the parent of the fava bean has yet to be found.[2] The native wild bean is believed extinct; maybe this should not be a surprise.[3] I imagine that countless generations of farmers selected the best beans for the following year's crop and would have removed less prolific wild varieties because they would have been prone to shattering, an event where the seed pod bursts open when the beans are dry and ripe – something farmers wanted to breed out of the domesticated crop so they could harvest the pods intact. Perhaps one day an intrepid plant hunter will chance upon an isolated population of wild fava bean in the uninhabited badlands of northern Syria, Anatolia or Iraq.

The debate about just when and where the fava or broad bean started life is far from over. Multiple domestications could have taken place over millennia, including in South Asia.[4] Over the thousands of years that this amazing pulse has been cultivated, there will have been countless local adaptions through domestication. Palmyra was founded more than 2,300 years ago and I have no doubt that the bean I was given had been grown locally for countless generations. The natural philosopher Pliny the Elder (23–79 CE) described the richness of the soil in the surrounding oasis and, with no evidence to back me up, I like to think that the large bean I brought home could be one of the oldest varieties under cultivation. Its short, fat pods, yielding a maximum of three seeds, show similar traits to other known ancient varieties.

Over centuries of breeding, farmers would have selected seed from larger pods with more seed in them for both yield and culinary worth. The short-podded fava bean is also known as the field bean. It is grown in the U.K. primarily as a fodder crop and for export to the Middle East for human consumption. The seeds are also much smaller than the one I was given in Palmyra. In the U.K., all types of fava beans eaten fresh are called broad beans. Exactly why is etymologically unclear. The Latin for bean is *faba* and in Italian, *fava*, which was anglicised to mean 'broad' in 1896.[5] Many broad beans have longer pods with up to eight seeds – sometimes more. I was to come across a wonderful example of this type later in my travels across Syria.

I can only hope that the resilience of those Palmyra farmers who remained and survived the catastrophe of the civil war will have enabled them to continue to grow and harvest this remarkable bean. At least it is safe in my little corner of Wales and is also now being grown by a few Syrian refugees in Canada, parts of Western Europe and the U.K., for whom it has great cultural significance and is the basis of much fine Syrian cuisine.

Another Syrian Gem

Damascus claims to be the oldest continuously inhabited city in the world. Some historians might disagree, but it is hard to argue with the length of its immense history. Supposedly founded by Noah's grandsons, the first written evidence of its existence is from 5,000 years ago. The Egyptians named the settlement Dimashq, now the Arabic name for Damascus, 3,500 years ago. Wandering through the Al-Hamidiyah souk and the alleyways of Old Damascus was a joy. The narrow streets were noisy with traffic and kids pushing large carts piled high with baby broad beans, which are traditionally cooked and eaten whole just like French beans. It was in this labyrinth that I found a small store selling packets of locally bred seeds, as well as sacks of peas, beans and lentils for smallholders. Among these sacks were broad bean seed of the variety I was seeing sold on the streets. Needless to say, the shopkeeper happily sold me a few grams of seed, which I have been growing in considerable quantity since that trip in 2011. They are delicious, not only when harvested and eaten whole, but also shelled, as is the usual way in the U.K. As with the Palmyra bean, I am able to share this wonderful example of Syria's culture and food heritage with refugees keen to grow them for themselves.

A South-East Asian Treasure

While on a seed-hunting expedition to northeastern Myanmar in 2015, I met a lovely old lady running a seed shop in the town of Hsipaw, which straddles the road from Mandalay to China. Her store was full of surprises. Myanmar, formerly known as Burma, was colonised by the British who fought three wars there between 1824 and 1885, before finally taking full control of the country in 1886. Knowing this, and aware also of the British love of broad beans, I was very intrigued to see a small box containing around 100g (3½ oz) of almost black fava beans on one corner of the shop counter. The shopkeeper told me that the beans were from her own garden. Local farmers no longer grew the bean because it was out of fashion, and she did not intend to grow them again.

She was in the process of chucking them out and gladly gave them all to me. So, how did those broad beans get to Hsipaw? Were they an old English variety that had been brought to Burma by the British or were they a genuine Shan heirloom that had arrived either from India or China? A nice piece of genetic research for a PhD biology student maybe? My original thought, that this bean had a British provenance, is almost certainly wrong. Local varieties of fava beans were domesticated across much of eastern Asia in Roman times. For reasons I explain later it is far more likely that the Hsipaw bean is closely related to those growing across the border in China and India. The colour of the seed – black – led me to believe the variety could be very old indeed since the very first cultivated fava beans had both black and white seeds. Varieties developed in the last few hundred years are either off-white or green. I would need to grow them to find out.

I had no idea how many of the beans I brought home from Hsipaw would germinate and in the event only succeeded in getting about twenty to grow. The seeds were very old, hence their appearance, and had been kept on a shelf where they were exposed to great temperature changes – from boiling hot in summer to freezing cold in winter – so it was a miracle any germinated at all. After a few weeks the short plants produced a profusion of pretty, purple flowers followed by short, stubby pods containing three seeds each. These were bright green and dried to a wonderful olive colour. Old seeds that have been stored for many years often darken with age. So, my Myanmar beans are not an ancient variety, but are certainly a genuine Burmese heirloom.

The beans were very tasty. Like their distant relative from Palmyra, I presumed them to be a very old heirloom, originally grown to be dried. With its continued cultivation in jeopardy, my serendipitous visit to that little shop in Hsipaw means the long-term survival of this bean is assured. I shall return one day with fresh seed in the hope that a local grower will want to bring them back into cultivation so that they can be part of the revival of traditional Shan food culture.

An Ancient Obsession

What I got so excited about when I came across that hidden gem in a dusty border town in northern Myanmar is how the fava bean has colonised the world. This is thanks in large measure to Neolithic man's skills as a plant breeder. The inexorable movement of the first farmers as they explored and settled in new lands, taking their seeds with them, meant the fava bean became part of a subsistence diet across Europe from the time of first settlement.

Egyptian priests considered it a crime to even look at a fava bean, believing the offending pulse to be unclean. Yet, the beans couldn't have been all bad because they have been found in Egyptian tombs and were the essential stuff of life for later civilisations. Ezekiel, in his Old Testament prophesies (Ezekiel 4:9), includes fava beans in his instructions for making bread, sufficient to feed oneself for 390 days! Pythagoras (570–490 BCE), despite being a vegetarian, forbade his followers from eating beans. He feared that the transmigration of souls could take one from being reborn, not as an animal or a human, but as a bean. His fellow Athenians, however, didn't all share his view of fava beans, including them in feasts dedicated to the god Apollo. Pliny the Elder waxed lyrical about the farms of Syria and also had a lot to say about the fava bean. Being a keen agriculturalist, he considered sowing beans to be as good for the soil as manure. He recommended ploughing the crop back in just as it was about to flower. Beans, like all legumes, store nitrogen from the air in nodules on their roots. This makes them a vital part of crop rotation and a means of improving soil fertility. Today I, like many growers, cultivate fava beans as a cover crop or green manure, turning this back into the soil and thus locking in nitrogen for the following crop. Whether or not Pliny understood the chemical process involved, there is no doubt that he clearly recognised the benefit to soil fertility of growing beans. He considered the fava bean to be the finest of all legumes.

Fava beans were used in soup and salads, possibly even re-fried. Seeds were split and roasted as snacks, just as they are today. The Romans also included them in sacrifices to the goddess Carna, protector of the human heart and other vital organs.[6] Her feast day, 1 June, happened at the time of the bean harvest. Beans were served mashed in lard or bacon: a dish still popular in Italy that goes back two millennia. And delicious it can be too. The Romans also believed that after death our souls take up home in beans, which made them an essential part of a funeral feast. There was one other role for the fava bean in Roman society: as a means of voting or passing judgement – a white bean in favour and a black bean against. In 1920 T.W. Sanders wrote in *Vegetables and Their Cultivation* that it was no doubt from this custom the modern term of blackballing undesirables arose. This refers to the habit of members of clubs dropping a black ball into an urn as a vote against in a secret ballot.

Deadly When Raw

Stories of our attitudes and beliefs about vegetables date back to the earliest written records – with many of them coming with a health warning. The fava bean was no exception since it can be highly toxic, even fatal. This threat to our well-being is due to the fact that the beans contain oxidants which can trigger a disease known as favism. It is still common today in parts of Africa, the Middle East, the Mediterranean and Asia. Globally, about 30 million people a year suffer from favism. Those of us with a particular genetic trait, meaning we have a deficiency in an enzyme known as G6PD, only need to inhale the pollen, let alone eat the beans, to suffer haemolysis, which is the rupturing of red blood cells. The disease especially affects children.* One can speculate that maybe Pythagoras had the G6PD mutation. Eating beans – even sniffing the flowers – would have made him very ill and might, in fact, have been the indirect cause of his untimely death. Fearful of being murdered while pursued by his enemies, Pythagoras had to make a ghastly choice: escape by running through a bean field or face his adversaries. He chose the latter option, was cornered and assassinated.

To make fava beans more palatable and less toxic, they must be well cooked, as boiling neutralises the G6PD enzyme and other toxins. Also, selection over thousands of years for beans with reduced levels of toxicity ensures that today's varieties are not going to kill us. To avoid a possible stomach ache, however, it is advisable not to eat the seeds raw unless they are very immature. This can be a test of resolve for a bean-lover like me, for whom freshly picked raw young

---

* The genetic recessive disorder known as Glucose-6-Phosphate Dehydronaise Deficiency (G6PDD) which causes favism is the result of an evolutionary development in some humans for resistance to some types of malaria, which is why it is most commonly prevalent in people from countries where this disease is endemic.

beans are utterly irresistible when I make my early summer forays into the vegetable garden.

Taking Over the World

The fava bean travelled far and wide and was introduced to the New World by Christopher Columbus who used it to provision his ships. (He also brought the pea for the same reason.) As I have found in my travels, this wonderful bean thrives in other countries beyond South America too. In China, the bean is a popular ingredient in Sichuan cuisine. Known locally as the Sichuan bean, it is the basis for a fermented paste called doubanjiang. It is also popular in the northeastern Indian state of Manipur where it is a traditional ingredient of the ethnic Manipuri people. This region borders northwest Myanmar and Sichuan is very close to Myanmar's eastern border. People of this mountainous corner of the world have grown the fava bean for generations, and it is likely that the region was also a place of first domestication.

In Britain the broad bean is the first pulse to be harvested in late spring and early summer. Being very hardy it can overwinter in most parts of the U.K. It has a long and rich European cultural heritage, having first been recorded on these shores 2,000 years ago by the Romans. There is still much debate about just how quickly it travelled west from the Fertile Crescent. One should not forget that Neolithic Man was busy building and living a settled life in western Europe over 5,000 years ago, so there is every reason to think that broad beans were on the menu when Stonehenge was being built, although archaeological evidence has yet to determine this. The Romans may have brought us many culinary delights, but the broad bean had been resident in

Britain for thousands of years before they arrived. Being hardy and suitable for storing, often for years, it was one of the most important sources of protein across Europe and important in mitigating famine. There is no written record of the cultivation of fava beans in Europe after the fall of the Roman Empire in the middle of the sixth century CE. However, our ancestors subsisted on a fairly miserable diet and fava beans would have been a mainstay: for so ubiquitous a food no comment was required! The arrival, in the sixteenth century, of the fava bean's South American cousin, the common bean, marked a slow decline in its importance for human consumption. The new bean became a fashionable garden crop – it was a very useful store of protein and did the same job as the fava bean. It was also colourful and delicious, but that's for another chapter.

From Field to Plate

Fava beans were considered an agricultural product and didn't appear in garden catalogues until the middle of the seventeenth century. They were sold not by weight but by volume. Years ago, I would visit the seed shop and buy my beans and peas by the pint or quart. Beans for growing in the vegetable garden started to appear in catalogues in the U.K. in the eighteenth century and varieties included Spanish, Windsor and Sandwich beans. Stephen Garraway, Seedsman and Net Maker in Fleet Street, London, listed ten varieties in his catalogues later in the century: Early Barbary or Mazagan, Lisbon, Broad Spanish, Long Podded, Nonpareil, Mumford, White Blossom, Green, Sandwich, Toker and Windsor. This list hardly changed for nearly a century and the names of these beans tell of their provenance, as they were clearly being grown throughout southern Europe and north Africa. These were all varieties that were grown to be eaten fresh too.

The widespread breeding of broad beans to be eaten young and freshly shelled is a relatively recent development in Western Europe. These types were originally known in the U.K. as Windsor beans before coming under the generic description of broad bean towards the end of the nineteenth century. Those grown to be dried continue to be known as fava beans, field beans or horse beans – because they were and still are grown as a fodder crop. In fact, both types are fully interchangeable for culinary purposes. I find it inconceivable that the earliest farmers would not have eaten some fava beans fresh, even though the crop was intended to be stored. Today, China is the largest producer in the world and Britain exports about 150,000 tons per annum to Egypt where demand outstrips production.

A Bean Renaissance

Of all the vegetables I grow the broad bean is probably the one that has had the least effort put into developing new cultivars. Show varieties, first bred by enthusiastic gardeners for length and number of seeds, became very popular in the nineteenth century before being taken up by commercial seed companies. There are two distinct types: hardy ones with kidney-shaped seed in long, slender pods that fall under the heading longpod – these are usually sown in the autumn – and rounder seed types in short, fatter pods known as Windsor. Many favourites still being grown and sold today have a long heritage. Green Windsor was introduced in 1809 and Bunyard Exhibition in 1884. A favourite overwintering longpod called Aquadulce Claudia – which continues to be the most popular bean for sowing in the autumn – was first sold in 1885 and White Windsor in 1895. All these varieties spawned very similar new ones, differentiated more by their name than their traits. In their 1939 catalogue, the English seed merchant Carters listed Mammoth Windsor, Colossal Windsor, Green Leviathan, Mammoth Longpod and Green Longpod. Top prize for a lack of originality in naming must go to bean breeders, I think. Things haven't changed much since. In 1950, Read and Hann of Hounslow, Middlesex, listed Eclipse (now sadly lost), Johnson's Wonderful Longpod, Taylor's Broad Windsor and Seville Longpod, all varieties one could have bought 200 years earlier.

The broad bean has largely been ignored by breeders in creating new F1 hybrids; just seven are listed in the 2021 European General Catalogue of Vegetables out of a total of 115 different varieties. Most of the modern cultivars in the list have been bred in the U.K., Spain, Italy and the Netherlands. Today's seed catalogues include some more recently developed cultivars. Thankfully, they continue to sell old favourites that are hard to beat for flavour and yield.

The Heritage Seed Library holds 41 varieties of broad bean, the majority originating in the U.K. and most with very local provenance. Others come from Ireland, the USA, the Netherlands, Greece and Eastern Europe. This is only scratching the surface. As a great lover of freshly harvested, young and lightly steamed beans, they are a welcome occupant of my vegetable garden. They may be an easy and forgiving crop, but they are, like their distant relative the runner bean, highly promiscuous, readily crossing with neighbouring varieties of the same species. So, to ensure that the seed I save is true, I have to both isolate the crops from each other and also be sure that my neighbours are not growing different varieties. I am fortunate that those of my neighbours who grow them are happy to sow the same varieties as me, especially when I provide them with seed.

Apart from the two Syrian varieties in my library I have an English favourite that, like so many of the crops I grow, comes from the Heritage Seed Library. Bowland Beauty can overwinter if grown in a polytunnel but is happiest sown outside in late winter or early spring. Much as I love all the beans I grow and eat, Bowland Beauty is hard to beat for its wonderful flavour and heavy crops of long, well-filled pods of lovely green beans. From its name I imagined it might first have been bred in northwest England, but I was wrong. Yorkshireman George Bowland, a farmer who had grown them for nearly 40 years, was the breeder. A true British heirloom.

There is another broad bean with a deep local association that I am glad to have growing from time to time. My Catalan kindred spirit Jesus Vargas gave me a very pretty and equally delicious broad bean that was grown in his neck of the woods, but which is now incredibly rare. Fava *Mourda Reina Mouz* – in English, Purple Queen – is highly endangered but survives thanks to a small number of enthusiasts in Catalonia. It's a pretty plant with lovely, fragrant, purple flowers and long pods with beans that turn purple as they swell. It is one of the great joys in my life to sow these bean seeds on a damp, dull and uninviting November day. Patience is required as they can take a month to germinate. But once up, they are determined things. The long nights and sharp frosts of winter deter them not. They may be slow to start but, as the days begin to lengthen in late January, they really get going and, with luck, by mid-March they will be a couple of feet tall and already sporting flower buds. As soon as the sun comes out and the air feels a little warmer, as if by magic, the familiar drone of large, white-tailed female bumblebees fills the air as they come to gorge themselves on the newly opened flowers. Most of my overwintering crops are grown in a polytunnel. If the bees fail to find the exit, I rush around after them with a jam jar and piece of card, so I can save them to continue their foraging unharmed.

Although I was well aware of the fava bean being a European introduction to South America over 500 years ago and thereafter widely used in Peruvian cuisine, little did I expect to find it growing in the driest place on Earth after Antarctica. It was February 2020, high summer in the Atacama Desert of northern Chile. The sun shines from a deep blue sky every day but, at an elevation of nearly 4,000m (12,123ft), daytime temperatures reach only the low 20s Celsius (high 60s Fahrenheit) and at night, more often than not, there is a frost. But even in this most inhospitable of landscapes, people have been farming for at least 3,000 years. Although most of the ancient terraces of the village of Socaire had been abandoned years ago, some were green with vegetation. It was while strolling among them that I found a tall broad bean. The crop of short, fat pods was heavy on the stems. From the oases and irrigated fields of the Middle East to the rich and damp fields of East Anglia, from the wet and humid mountains of the southern Himalayas to the driest desert in the world, fava beans had made a home. A remarkable testament to the resourcefulness and genius of farmers both ancient and modern, but also to the ability of one of our oldest domesticated foods to adapt and find a home almost anywhere.

Orange is Not the Only Colour

Never bolt your door with a boiled carrot

Irish proverb

It's an unseasonably warm mid-April morning and one of the great pleasures in my life is waiting for me in the polytunnel. Her billowing fronds, their greenness as bright as a go-sign on a dark night, are saying: 'Pull me, pull me.' And who am I to refuse such a compelling request? I know for sure what delights are hidden below the soil and that my taste buds are in for both an unforgettable experience and a reminder of just why I have loved growing this particular vegetable since my first days gardening. One of my first, the carrot has remained a perennial on my plot for as long as I can remember and for a very good reason. Freshly pulled on a spring morning from damp, warm soil, its heavenly perfume filling the air, that orange root begs to be wiped clean on the back of my trousers or given a swift rinse under the outside tap and then immediately consumed. The sweet perfection of a traditional early variety like Amsterdam Forcing is unforgettable.

Not as Old as You Think

The domesticated carrot started its life in Afghanistan where the Hindu Kush meets the Himalayas. The early domestications had branched roots, which did not store well and were prone to bolting (premature flowering). Purple or even deep red in colour due to the presence of the compound anthocyanin, these early domestications turned brown on cooking and their roots, which were somewhat inedible, would have been used as a dye as well as a medicine. The domesticated carrot we eat today is considered to have been first grown across a wide region from Anatolia in the eastern Mediterranean, throughout Iran, Pakistan, southern Russia and as far east as northern India. It was brought to western Europe about 800 years ago, as we will discover later.[1]

The wild carrot I find growing at home is a close relative of the domesticated carrot, indigenous to Europe and parts of central Asia, and has been used as an herbal remedy for thousands of years. It is commonly known in the U.K. as Queen Anne's lace because of the very pretty, lace-like appearance of the creamy flower head. It has a red centre which is said to represent a drop of the queen's blood. This was allegedly shed when, during Anne's reign between 1702 and 1707, she pricked herself with a needle while embroidering. Having this wild ancestor growing nearby, however, is a nightmare for seed savers like me because it will readily cross with its cultivated cousin. Nevertheless, the flowering carrot is a wonderful thing with a heady perfume and the ripening seedhead a beautiful example of nature as an abstract artist. As the flowers set seed, the umbels turn up at the edges and darken to shades of brown, taking on the appearance of a bird's nest.

Anyone who has tried to eat the root of a wild carrot will know that it is not a pleasant experience. Although it has a carrot taste, it is tough and its real value comes when put in the hands of herbalists who use the whole plant, dried and ground up, for a multitude of ailments. The plant has generally been used as a diuretic, stimulating the flow of urine. The seeds are antilithic – that is, they prevent the production of uric acid – and so were used as a remedy for gout. Carrot seeds are also alleged to be good at reducing flatulence and calming the stomach. For some it was, apparently, also an antidote to snake bite. The phallic appearance of the humble carrot has been credited with giving it great aphrodisiac properties too (why is it that any edible plant that faintly resembles a penis is believed to be good for one's sex life?). In fact, carrot seed, which contains small amounts of oestrogen, has been used at the very early onset of pregnancy as an alternative to the morning-after pill. Traditionally, a tincture was taken, or the seeds chewed a week or so before and after ovulation as a form of birth control. Eating the seeds was also believed to help with conception! Herbalists over the centuries have attributed conflicting medicinal qualities to most vegetables that have been part of their arsenal of cures. Carrots are no different but, because the seed can induce uterine contractions, pregnant women should never eat them.

Further Back in Time

There is archaeological evidence of purple as well as yellow carrots having been cultivated in Afghanistan some 5,000 years ago and there would also have been white and even black carrots too, thanks to the genetic diversity of early domestications throwing up colourful mutations. Carrot-like plants have been found in Egyptian drawings from 2000 BCE, but there is no evidence that the Egyptians were carrot eaters at all, despite what some Egyptologists claim. The only tangible evidence for this would be the presence of carrot seed at archaeological sites and of that there is none. But, as wild carrots have been identified along much of the North African coast, one can only speculate as to whether carrots were ever part of a pharaoh's diet.

There is evidence that wild carrots were cultivated by the Romans and, apparently, much loved by the emperor Tiberius. Their principal value was pharmacological. The leaves and flowers are aromatic and so made a useful addition to a salad, but it was really the seeds (and possibly only later the root) that herbalists used the most. It wasn't until the Middle Ages that carrots were first bred exclusively to be eaten as a root vegetable.

It is difficult to be certain whether the Greeks and Romans truly enjoyed carrots because the wild carrot was often mistaken for another native wild root, the parsnip (both being white, after which the similarity ends). The parsnip and carrot were a part of the Greek and Roman obsession with health and sexual capabilities, although they did not differentiate between them. The two plants were sometimes called *Pastinaca*, which is derived from the Latin for 'to dig', which was of no help in bringing clarity to the carrot story. It wasn't until the second century CE that the Romans distinguished the wild parsnip from the wild carrot, calling it *carota*, and this is the name that has stuck ever since. One of the names given to carrots by the Greeks was *Daukos*, which was adopted by Carl Linnaeus (1707–1778) who combined the Greek and Latin names to give us *Daucus carota*. *Pastinaca* he reserved for the parsnip. The English word carrot derives from the French *carrotte*. In the eighth century, King Charlemagne mentions carrots as a useful crop, principally as animal feed, although unlike the Arab agriculturalist Ibn al-'Awwām writing some 400 years later in Spain, he does not refer to it as being a domesticated type. Of its colour we know nothing, but it was most likely either white or purple. Although there is no written evidence of their precise diet, there is an academic assumption that cultivated strains of the wild parsnip and carrot were an important part of a peasant's diet from Roman times until the seventeenth century. Carrots could be lifted and stored through the lean winter months and parsnips were hardy enough to be left in the ground and lifted as required. Today's modern farmers leave overwintering carrots in the ground covered in straw and polythene, ready to be harvested mechanically according to market demand.

I began the chapter by waxing lyrical about the deliciousness of this humble root. Yet, its culinary qualities are the result of no more than about 500 years of selective breeding. Since the wild ancestor of the carrot is so widely distributed across Europe and Asia Minor, it has been a challenge to fully understand the process of domestication. This is because there is no evidence that it simply evolved from the wild carrot. Furthermore, breeders have been unable to develop an edible root just from the wild carrot. Nikolai Vavilov (1887–1943) described the wild parents of the first domesticated carrot. During his travels across Russia and its eastern neighbour, researching his theory on the centres of origin of cultivated plants,[2] he identified two sub-species. One was found growing on the Iranian plateau. It is this carrot that gave us purple and yellow varieties. The second sub-species was white and is indigenous to Turkey. Its domestication gave us the yellow and later the ubiquitous orange carrot.[3] Both these wild species would have accidentally hybridised and mutated, and it would have been these new forms that farmers first selected and, through continuous selection over many generations, bred to produce varieties similar to those we enjoy today.

From earliest times the carrot has been grown to feed livestock through the winter. This tradition continues, as a visit to a farm supply store today will attest. Nets of scrubbed carrots intended for pets and horses are usually on sale. They are just as good – and tasteless - as any shop-bought ones. Sadly, even today there are those who eschew this wonderful root; something I find hard to believe. Also, for reasons we do not yet fully understand, some people can have an allergic reaction to carrots, especially when they are eaten raw, which in some circumstances can prove fatal.

It's All About the Colour

We have the Arabs to thank for introducing today's carrot to Western Europe. As we have seen, there are two distinct sub-species that led to the domesticated carrot. The sub-species *sativus*, native to Turkey, was grown by the Arabs and much enjoyed by their invading armies, both animal and human. Over a thousand years ago, at the end of the tenth century, carrots are mentioned in a cookery book complied by Ibn Sayyār al-Warrā, an author from Baghdad. Called *Kitab al-T. abīh˘* (Book of Dishes), I imagine the book being added to the libraries of Europe's Moorish invaders who had started their own vegetable gardens in the Iberian Peninsula early in the eighth century. The first historical record of carrots as a crop in Spain and southern Europe, however, is found in the work of the great Arab agriculturalist Ibn al-ʿAwwām, towards the end of the twelfth century. It seems that by this time there were a number of different but unnamed varieties of carrot being grown.

Carrots came into cultivation in northern Europe some 200 years later and it would appear that they were valued for their high sugar content – recipes of the time have them being turned into jams, sweet condiments and puddings. Although they came in a variety of colours and shades – red, white and yellow – it was the yellow ones that became the most favoured in Europe because they were sweeter and didn't turn a muddy brown when cooked. Carrot colour has been the subject of much scholarly discourse over the years and, whether the orange carrot existed before the attentions of Dutch breeders is explored later, so I use the word 'yellow' with some literary licence.

While Moorish invaders were introducing southern Europe to the western sub-species, *sativus*, its relative *atro-rubens* spread further east from Iran and the Hindu Kush along the Silk Road. Modern genetic sequencing points to the fact that Chinese carrots, which come in red, white, purple and orange, are all derived from *atrorubens*. Similarly, deep red descendants of this branch of the family remain firmly part of the food culture of Rajasthan, a state in northeast India that was once controlled by the Moguls. The homeland of these Muslim descendants of Genghis Khan was in the heartland of the carrot's birthplace. Coloured varieties are now becoming trendy in Western food culture, having been a staple in the East for centuries.

On a seed-hunting trip to India in 2019 I was able to enjoy freshly harvested carrots in the same way as if I had been back home in my own garden. The location was a small village, Jaisinghpura, half an hour's drive southwest of the city of Jaipur in central Rajasthan on the eastern edge of the mighty Thar Desert. I had been hanging out with a bunch of farmers, all making a living from the land growing several *desi* (local) varieties of vegetables. One of them was Ramgilal. With his toddler son in his arms, he proudly gave me a guided tour of the 30-acre (12-hectare) farm he shares with his two brothers. We were in his carrot patch, and he pulled a long, large, deep red specimen from the sandy soil. I wiped it on the seat of my trousers and bit into the sweet, crunchy flesh. This is a carrot I had seen being sold in markets everywhere on my travels through the state. It's the one everyone eats, a mainstay of Rajasthan's indigenous food culture. It looks spectacular, a beacon on any market stall, and eaten freshly pulled, it was something divine.

The Colour Orange

Columbus's arrival in the Caribbean in 1492 sparked a transfer of native vegetables in both directions across the Atlantic. Carrots could be stored for long voyages and were planted by the colonisers who followed him. However, it was not until the beginning of the seventeenth century that the carrot was to undergo a dramatic change of fortune, in more ways than one. As the sixteenth century drew to a close, Flemish growers started to work on improving the colour, yield and appearance of the carrot as well as its eating quality. Yellow, western varieties, being biennial, were not only less likely to bolt than their eastern cousins, but also genetically predisposed to grow a single bulbous root, full of sugars and flavours. The darker the yellow, the more breeders liked them.*

The word 'orange' is relatively new to the English language and first appeared in a reference to clothing belonging to the Scottish Queen, Margaret Tudor, in 1502.[4] The orange, which is native to China, arrived in Europe with the Arabs at the beginning of the eighth century and was called the *sinaasappel* (Chinese apple) in Dutch. The Spanish took the Persian word for the fruit, *narang*, referring to the bitterness of its skin, and called it *naranja*, which in Old French translated as 'orange'. The 2011 edition of the Oxford English Dictionary describes the colour orange used in Old English as *g.eolurēad* (yellow-red). This name for the fruit was probably adopted into Middle English at the same time as the orange first appeared in Britain after the Norman Conquest in 1066, but it was not used to describe the colour of a carrot until much later. So, it is not surprising that descriptions of carrots of all shades of yellow and red didn't describe them as 'orange' until the word became a common adjective in sixteenth-century English. Because of this, earlier descriptions fail to help the researcher achieve clarity as to a variety's true colour.

Although principally red carrots were being cultivated across Europe from the eleventh century, it was thanks primarily to

---

* The red carrots, which are descended from the Eastern parent I enjoyed in Rajasthan, are annuals. A few prime specimens are left to go to seed after the crop is harvested. Western carrots are biennial, the result of domestic selection, which allowed farmers to lift and store them through the winter. Selected roots would then be re-planted in the spring to be allowed to go to seed. Carrots are not the only biennial grown by gardeners. Many other root crops like beetroot and parsnip are also biennial, as are onions and some brassicas.

highly selective breeding by the Dutch that 500 years later the orange carrot we recognise today became ubiquitous. As a kid I was told that the orange carrot was a symbol of the Dutch Royal House of Orange. It was used as a propaganda tool when William and Mary took over the British throne after a bloodless coup in 1688 known as the Glorious Revolution. William inherited the title of sovereign Prince of Orange after a feudal principality, complete with orange groves, in Provence, southern France. Sadly, the story that breeders created an orange carrot to honour the Dutch royal family is pure myth, but since when has fact been allowed to interfere with political propaganda? The reality is that the Dutch were growing orange carrots long before William inherited his title and moved to England. But the orange carrot is the national vegetable of the Netherlands and many of its people still cling to the idea that its colour was created as a tribute to the House of Orange. As a marketing strategy and way to raise 'brand awareness', it was brilliant, and I think it would be churlish to disabuse them of their belief. Also, now that the genome of the carrot has been unravelled, we know that orange carrots are the direct descendants of yellow varieties and are testament indeed to the genius of Dutch breeders.

A Long-Lasting Heritage

Carrots had become part of a subsistence diet throughout Europe and the Americas by the seventeenth century, but different varieties were yet to be given names. A seed seller from London, William Lucas, lists red, orange and yellow carrots in his catalogue of 1677 and, although Dutch breeders had named varieties, these were not shared with consumers for another hundred years. At the end of the eighteenth century, English merchants at last listed a few named varieties. The Curtis seed catalogue of 1774 includes three: Early Horn, Short Orange and Long Orange. In 1780, J. Gordon of Fenchurch Street lists just two carrots: Early Horn and the Orange or Sandwich carrot (Sandwich refers to where they were grown). Carrots like to grow in a light soil, and Sandwich in Kent fitted the bill perfectly. Flemish immigrants escaping Catholic persecution in the latter half of the sixteenth century had settled there and grew them, including for their new Protestant queen, Elizabeth I. We also know that Early Horn is one of the oldest named varieties and is related to many of those we enjoy today.

Not only do we have to thank Dutch breeders for the ubiquity of the orange carrot, but it is also to a Dutchman, O. Banga, writing in the early 1960s, that we should give thanks for a considerable body of work on the history of carrot cultivation and breeding.[5] He identified two Dutch varieties, Scarlet Horn and Long Orange, as being the progenitors of pretty much all of today's orange carrots.

Through genetic analysis we now know those purple carrots that originated in Afghanistan mutated into yellow ones. Also, we need to remember that descriptions of carrots as being red actually describes those coloured purple – think of red cabbage and red beetroot. Colour changes of the earliest cultivated carrots happened through accidental mutation, rather than hybridisation. The Western Europeans' preference for the yellow over the purple carrot was encouragement enough for those eighteenth-century Dutch breeders to work on ever-deeper yellows until they got a sweet and tasty orange the consumer would buy. By the middle of the eighteenth century, we had new varieties: Early Half Long Horn, Late Half Long Horn, Early Short Horn and Round Yellow; the last two being the parents of nineteenth-century classics, Paris Market and one of my favourites, Amsterdam Forcing. It is testament indeed to the quality and skills of breeders that these two early varieties continue to be hugely popular after over 250 years in cultivation. Other carrots such as Nantes types – those with cylindrical roots – were the result of a century of breeding from the now extinct cultivars Late Half Long Horn and Early Half Long Horn. The name suggests the French had a hand in developing this type. Early twentieth-century breeders, according to Banga, gave us Imperator – a long, tapering type, which is a cross between the Nantes and Chantenay, a red-cored variety (delicious by the way) that had been bred from another eighteenth-century variety called Oxheart. Imperator types are the basis for most modern cultivars developed for today's supermarket trade.

One variety that I grow every year is Autumn King, an open-pollinated stalwart that has been around for a century or more and, thanks to climate change, one that can sit happily in the soil through the winter to be harvested as and when required. The days of clamping – storing carrots in mounds of sand – are well and truly over, for me at least. The prettily named Flakkee, a very good overwintering storage variety, has claims to Italian heritage. It is synonymous with Autumn King, which begs the question: do we have here another example of breeders renaming varieties to suit their own markets and cultural sensibilities? Fortunately, many of these very earliest breeds of carrot are still with us and, regardless of what they are called, they are a culinary delight.

Ploughing my way through the 2015 EU Common Cat-alogue of Vegetable Varieties, I counted 629 named culinary carrots, of which 332 were F1 hybrids. Pretty much every commercial colourful carrot, including orange ones grown today, is a modern hybrid. Gardening catalogues are filled with F1 hybrids as well as traditional open-pollinated varieties, including a new generation of cultivars with resistance to the scourge of all carrot-growers: the dreadful carrot root fly.

A Dreaded Fly

The traditional domesticated carrot is notoriously susceptible to carrot root fly, a nasty insect that can ruin a crop. Fortunately, as we shall see, modern plant breeders have been able to come up with a solution. Because domestication came about through the selection of mutations of wild carrots as well as through hybridisation, wild and cultivated carrots can easily cross, and this has been exploited by plant breeders to produce new fly-resistant cultivars. The wild species *Daucus capillifolius*, which was identified by the botanist Elfrid Gerhart in sandy soil near Tripoli, in Libya, in 1956, shows a fantastic resistance to carrot root fly because it contains very low levels of chlorogenic acid, a chemical which carrot-root fly larvae need to survive. It would cross readily with the cultivated carrot and was the subject of research undertaken at Wellesbourne, Warwickshire, in the late 1980s and early 1990s.[6] Researchers created crosses and hybrids using a lovely old-fashioned variety, Danvers Half Long, named after Danvers in Massachusetts, USA, where it was first bred in 1871. The result has given us delicious resistant varieties such as Flyaway and Resistafly. Sadly, my favourite cultivar, Sytan, is no longer for sale. However, to be sure of a crop entirely free of carrot-root fly larvae, I always employ a physical barrier such as garden fleece staked around the crop. A number of other wild relatives have recently been identified and are now a vital part of the breeding of fly-resistant cultivars. As the public becomes ever more aware of the need for us to grow crops in a sustainable way with fewer chemical inputs, including pesticides to kill carrot-root fly larvae, being able to grow carrots with resistance to this insidious pest can only be good for the planet.

Purple Does Not Mean Heritage

One of the things that really gets up my nose is eating in a restaurant that has 'heritage carrots' on its menu. This is marketing nonsense. These carrots are for the most part purple, yellow or white, which apparently allows the chef to call them heritage, but they are, in fact, mostly modern hybrids. Unless they are named and have provenance, these roots are fakes. However, there is one stalwart heritage variety you won't find on the menu, and that is Red Elephant. I always make sure I have a good supply of seed of this wonderful variety. The British seed company Carters waxed lyrical about it in its catalogues in the 1930s, describing it as 'A veritable giant both in length and bulk; specimens have been exhibited measuring 30 inches [65 centimetres] in length; a remarkable variety prominent in the garden and on the exhibition table. 141 first prizes reported by customers in one season!' Until recently it could be bought in Canada and New Zealand. Now I must save seed myself to be able to enjoy its wonderful flavour. It was first bred in Australia in the nineteenth century, but production was discontinued there in 1910. Thankfully, this magical carrot was a horticultural export, hence its inclusion in the Carters' catalogue for over a decade. I have no idea why the carrot was discontinued in Australia, and I would love to get it back into the local diet because it is a very fine carrot indeed. So far, all my efforts to achieve this have failed – understandably, Australian plant quarantine authorities are wary of foreign plant material and even importing seeds of a native variety requires much bureaucracy and cost. So far, Australia's loss is my gain and Red Elephant is the one carrot I grow out for seed every few years.

Hidden in Plain Sight

The story of Red Elephant is yet another example of how fragile and tenuous the conservation of the genetic diversity of our crops can be. Maybe, in the case of the humble carrot, much less has been lost than with other food crops. Casting my eye through a seed catalogue today and comparing it with Carters from 1939, I can find, alongside modern varieties, including F1 cultivars bred for carrot-fly resistance, those that have been a part of European and American food culture for the last 400 years, with names such as Berlicum, Parisian (French Market), Danvers, Chantenay, Early Nantes Short Horn and Early Scarlet Horn. Despite remarkable advances in carrot breeding to deliver uniformity, as well as disease and pest resistance, the old ones are still the best and much loved. Perhaps the world's love of carrots and the importance of colour in different societies and food cultures means the many traditional varieties that have been grown for centuries will continue to thrive alongside modern cultivars which are the product of sophisticated modern plant breeding techniques. A future less certain awaits many of our other vegetables.

The carrot stores and travels well, yet just a few short hours after being harvested much of the richness of its wonderful flavour is lost. I suggest to anyone who is contemplating growing a few vegetables to find a corner of their garden for carrots because, once you have eaten a homegrown specimen, you will never want to buy a carrot again, for nothing beats the flavour of a freshly pulled carrot and that, dear reader, is a fact.

In Search of a
Welsh Leek

'Eat leeks in March and wild garlic in May,
and all the year after the physicians may play.'
Traditional Welsh rhyme

I t's late summer and the great British weather is doing
its worst. The rain is drumming a relentless riff on the
marquee roof as I squeeze between the somewhat damp
and steaming lines of admirers who are ogling the giant produce on show at an important cultural event: the annual Usk
Show. Gardeners have been competing ruthlessly against each
other since Victorian times to grow the longest, the heaviest,
the tallest, the fattest, the most perfect-looking giants of the
kitchen garden. There is a whole subculture of horticulturalists
whose life is dedicated to growing exhibition crops in the hope
of winning a prize, the admiration of an appreciative audience
and the envy of their fellow exhibitors. I am a reluctant competitor for two reasons. Firstly, the idea of someone growing
a vegetable judged better than mine is something I find hard
to bear. But secondly, and more importantly, exhibiting vegetables is all about appearance and has nothing to do with taste.
An utterly pointless pursuit in my opinion, but one gardener's
giant brag is another's culinary delight.

There is something incredibly British about growing leeks and it is at our county shows or vegetable society bashes that we see how huge these things can become. Just don't ever ask me to eat one. These leeks are different to those which the average gardener grows, not just because of their great length and girth, but also because they are not usually grown from seed. Instead, they are cultivated from offsets which are clones of the parent. Grown in this way, competition leeks are all genetically identical to each other and it is the skills of the grower that determine its perfection and not its personality on the plate, as flavour is not judged.

When it comes to enjoying the finest of culinary leeks, it is to the seed packet one turns. This hardy member of the same family as onions, shallots and garlic has been domesticated from the wild leek, *Allium ampeloprasum*. This name is derived from two words: *prason* is the Greek word for 'leek' and *ampelo* means 'vine'. The Latin name for the wild species means 'the allium that grows in the vineyard'. It was the Roman physician Dioscorides (*c.*40–*c.*90 CE), who noticed that vineyards were much favoured by the plant.

A Plant of Many Talents

The wild leek probably arrived in Britain in prehistoric times.[1] It grows mainly in parts of South West England and South Wales, on rocky coastlines and damp places that have undergone human disturbance, such as the building of drainage ditches. There is a story that the wild leek was brought to the island of Flat Holm, a small blob in the Bristol Channel, by Augustine monks in the twelfth century. However, it is more likely, as we shall see, that the monks brought with them the seed of a cultivated leek which would then have crossed freely with any wild leeks on the island.

Wild leeks are also found in the northwest corner of Wales, at South Slack on the island of Anglesey, where they had a particularly good year in 2013 when the pretty flower spikes reached 2.5m (8ft) in length. The wild leek would have been used primarily as a tonic in ancient times – it may be less effective for its medicinal qualities than its relative, garlic, although some herbalists dispute this. Nonetheless, wild leeks, which could apparently deter moles, were an important cure for the common cold and as a tonic to restore the iron lost after menstruation. Leeks were also considered a useful aid for childbirth and for avoiding being struck by lightning. They were used as a moth and insect repellent too, but I imagine smearing leek juice all over oneself to keep bugs at bay might have proved a human repellent as well. Today, a cursory trawl of the internet reveals plenty of other claims for this vegetable. It is said to be good for the heart, to lower blood pressure, to have powerful anti-cancer properties, be good for brain function, and reduce the chance of having a stroke. How can we argue against eating leeks when they are obviously so good for us? The leek is also not alone among foods that have a phallic appearance, having been associated with love and lechery – a great vegetable to pep up one's love life.[2]

Emblem of National Identity

The leek is one of two great Welsh cultural icons: the other being the daffodil whose Welsh name, *Cennin Pedr*, means 'Peter's Leek'. The leek received a special place in Welsh culture and British literature at the Battle of Crécy in 1346.* Colourfully described by the Bard himself in his play *Henry V*, written 250 years after the battle, the comical soldier Fluellin tells the king about Welsh soldiers wearing leeks in their Monmouth caps. Early in the twentieth century there was much ill-tempered hot air, mudslinging and verbiage written by some patriotic Welsh botanists and local historians about which plant was the true emblem for Wales. Efforts to discredit earlier references in favour of the leek, including from Shakespeare, were legion. Needless to say, the daffodil mob poured scorn on all historical texts that cast doubt on their claim. And any reference to the Battle of Crécy was 'of no trustworthy authority'. In 1911, the British Prime minister David Lloyd George, who was a committed advocate for the daffodil, made sure that it was used as the official symbol during the investiture of the new Prince of Wales at Caernarfon Castle. The leek was nowhere to be seen; well and truly toppled from its emblematic pedestal. The battle for the leek may have been lost but the war was not over.

In 1919, a Fellow of the Linnean Society, Eleanor Vachell, wrote a paper for the Cardiff Naturalists' Society, titled *The Leek: The National Emblem of Wales*, where she attempted to put the record straight. It makes hilarious reading:

> '...no writer on this controversial subject seems to have viewed it from a fair standpoint and to have presented fairly and honestly the just claims of the rival plants, all having hurled epithets and

---

* Crécy is in northeastern France and the battle took place during the Hundred Years War between England and France. The French king at the time was Philip VI and England's was Edward III.

scorn upon the plant they do not favour; apparently believing in that way to injure its cause. Thus we find the daffodil referred to by its enemies as a "sickly and maudlin, sentimental flower, the favourite of flapperdom," while the leek is spoken of as an obnoxious or common and garden plant...it is greatly to be hoped that the evidence in favour of the leek will be considered sufficient to justify it being accepted as the true emblem of Wales, and to put an end to the ridiculous mixture of leeks and daffodils that are now worn by our countrymen on St. David's Day, making them the laughing stock of other nations...'[3]

A sense of national outrage persisted with fabulous accusations hurled indiscriminately by the proudest Welshmen at the despised English. A certain Mr Llewelyn Williams wrote, at the height of the controversy in the first decade of the twentieth century, suggesting that the mistake of substituting a 'stinking vegetable' for a 'charming flower' was 'due to a blunder made by Shakespeare or some other equally ignorant Saxon' who confused the Welsh *cennin*, meaning 'leek', with *Cennin Pedr*.

In Search of a Welsh Leek

Whether or not there is a true Welsh heritage or heirloom leek – something I have yet to find – or that the nation actually enjoys eating them, is merely a sideshow to the great cultural significance of the leek. Its appearance on St David's Day, on 1 March, celebrates a much earlier battle – that fought by King Cadwallon of Gwynedd sometime in the seventh century when his soldiers allegedly wore leeks in their helmets as identifiers. The legend tells how the plucky Welsh soldiers fought off the hated Saxon invaders, having helped themselves to the leeks growing in a nearby field. St David himself was said to have lived on a diet of bread and leeks, which ensured the saint had a sonorous and clear voice that enabled the multitude to hear him when he preached to the Welsh synod.

However, when looking at the two records of the wearing of leeks in battle, it seems that those arguing for the leek as a Welsh emblem displayed horticultural ignorance. The Battle of Crécy took place on 26 August 1346, when cultivated leeks would have been no thicker than a pencil and thus an insignificant emblem. Nonetheless, King Henry V's troops, we are expected to believe, had the opportunity to follow in the footsteps of their Celtic forebears and stick a leek in their caps. In early March, however, when King Cadwallon was at war, leeks would have looked magnificent! For the sake of botanical clarity, one should remember that the Welsh leek of the seventh century would almost certainly have been introduced by the Romans, whose love of all alliums, including leeks, is discussed later. It is most unlikely to have been the wild type which, even back then, was only found growing in rocky and remote corners of the west coast of Wales.

In her wonderful book *The Origin of Plants*, Maggie Campbell-Culver reminds us that the entire native population of Britain, following the Roman occupation, would not have made a distinction between the vegetable garden and farmland used for cultivating crops. Life was a matter of survival and the entire community would be focused on finding and growing enough food just to stay alive. Only monks and a few scholars could read and write, and even they spent most of their time growing and foraging. Unlike today's common expression of a kitchen garden as being 'the cabbage patch', at the time St David was battling the Saxon invaders the description of a place where vegetables were grown was the *Leac-garth* (herb garden), *Leac* meaning 'herb' in Anglo-Saxon. There would seem little doubt, therefore, that leeks were an important part of Anglo-Saxon food culture. Our obsession with this allium continues to this day, as is evident from its importance on the show-bench. Perhaps the literal translation of the location of the battle led by King Cadwallon – an unidentified field of leeks somewhere in South Wales – says more about the vegetable's place in Welsh culture than what actual species the Welsh and their hated Saxon enemies could have been trampling over when they helped themselves from some impoverished peasant's vegetable patch. In any event, the leeks those brave soldiers fighting for Henry V wore would have been a French variety, the result of a thousand years of cultivation after the fall of Rome and a distant relation of the common leek we enjoy today.

No Wild Side

Unlike many of the vegetable characters I write about, although domesticated from the wild leek, the cultivated leek is known botanically as a cultigen – a term used to describe a cultivated plant lacking wild counterparts. This is because domestication and selective breeding over many thousands of years has given the world a species that is now quite distinct from its wild ancestors.

Records indicate that the cultivated or common leek was domesticated in Egypt and Mesopotamia at least 4,500 years ago. However, as the wild leek is native to all the countries bordering the Mediterranean and the Black Sea, as well as

77

the western and northern parts of the Fertile Crescent, it would most likely have been foraged by our ancestors long after it was domesticated. It is considered naturalised in northern Europe, including parts of the U.K., throughout Indo-China and, as a result of European colonisation in the last 600 years, much of Australia and parts of the North and South American continents.

The long process of domestication has given us a number of distinct forms which have embedded themselves in several different food cultures. Wild leeks are onion-like in appearance, with small bulbs or clusters of bulbs at their base. Domesticated leeks are divided into five distinct groups, from which several different types were perfected: the Common Leek, the Greater-headed Garlic, the Pearl Onion, the Kurrat, and the Taree or Persian Leek.

It is the changes in the concentric, unsheathing leaf base that were selected by farmers to create long, edible pseudo-stems. These are false stems formed from the swollen leaf base which we associate with the distinctive appearance of the common leek.[4] Elephant garlic is a member of the Greater-headed Garlic group and those of us who have grown and eaten it will attest to it looking more like a giant garlic bulb than a type of leek. Only when eaten does its leek-like flavour come through, and those expecting the pungency of real garlic will be disappointed. Ditto a close Chinese relative, the Pearl Onion, which is known as a button baby onion in the U.K. and a creamer in the USA. Frequently sold as a pickled onion, it is, in fact, a type of pickled leek! The other two, the Kurrat and the Taree, look similar to the common leek. The former is the result of selection from forms of wild garlic, with an emphasis on the leaves, and remains popular in Egypt today. It's very much a 'cut-and-come-again' crop, harvested every few weeks until the plant is exhausted – which can take over a year. The Taree is very similar to the Kurrat, but with narrower leaves, and is a popular ingredient in the cuisine of northern Iran. Seed from the Greater-headed Garlic group is infertile, but the other four groups are interfertile and will readily cross with each other, producing hybrids of variable culinary worth.

There are other alliums called wild leeks, which are an entirely different species. I think of the American wild leek, *A. tricoccum*, commonly known as ramps, which is foraged in the spring and sold at farmers markets. It is rather yummy and, confusingly, looks like the unrelated wild garlic *A. ursinum* which I forage near my home. Also known as ramsons, wild garlic is a wonderful, sweet-smelling sight in springtime when great drifts can be seen in British woodland glades and along rural roadside verges.

Food Stories

Thousands of years before the Welsh started to squabble over the cultural significance and merit of the humble leek, it had cemented a place in the earliest recipes of the eastern Mediterranean. The stories of the leek's journey onto our plates are closely entwined with those of its two cousins, garlic and onion. They are referenced on Sumerian Cuneiform tablets, nearly 4,500 years old, along with other culinary essentials. Tablets from Mesopotamia, dating back to between 1600 and 1700 BCE, and known as the Yale Babylonian Tablets, probably represent the oldest cookery book in the world and include around 40 recipes in which the *Allium* genus figures large. On the tablets, the leek – most probably the Taree – is called *karsu*. It would appear that the civilisations of

Mesopotamia liked to cook leeks and garlic together purely for nourishment, as there is no record of them having any medicinal or magical properties. In the Old Testament there is a description of foodie pleasures lost to the Israelites who, while wandering the desert with Moses, thought of the good things they had left behind: '...We remember the fish which we did eat in Egypt, the cucumbers, melons, leeks, onions and garlic...' (Numbers 11:4–6).

It would appear that the Egyptians also enjoyed their leeks, which would have included the Kurrat, and even venerated them. They appear on hieroglyphics and on the head or in the hand of the god Osiris. The Roman emperor Nero (37–68 CE) may have died young, but I doubt this can be blamed on his love of leeks. Like St David, who believed eating leeks helped lubricate his larynx, Nero would regularly eat large quantities to help his singing voice. As a result, he was given the nickname *Porrophagus* – 'leek-eater' in Latin – an insulting term from an ungrateful empire. Like so many vegetables that are at the heart of European cuisine today thanks to the Romans, the domesticated leek was grown across their empire.

Anglo-Saxon Europe embraced the cultivated leek. Nutritious and a reliable, hardy crop, it could be harvested from late autumn to late spring and would have been part of a daily diet, which is why it has such an important place in our folklore and literature. Eating leeks also had negative connotations. A century or more ago, derogatory comments about people being 'as green as a leek', or something being 'not worth a leek', were a regular part of English vernacular.[5] As well as praising the Welsh in the Battle of Crécy, in one scene in *Henry V*, Shakespeare has the Welsh character Fluellen force his English compatriot Pistol to eat a leek because 'If you can mock a leek you can eat a leek.'

On a more positive note, it is not just the Welsh but also the Romanians who have the leek as a bedrock of their cultural identity; the Scots do, too. One just has to think of Cock-a-Leekie soup, a meal in itself, with many regional variations that all include leeks. The other quintessential Scottish dish is porridge, the name being a combination of the word 'pottage' and the French word *porée*. The Latin word for leek, *porrum*, became *poireau* in French and *Porree* or *Lauch* in German. *Porrum* was first used to describe a thick, leek-based vegetable soup like Cock-a-Leekie, which became extended to describe any glutinous and soupy dish made using dried peas or oatmeal.[6]

Into an Age of Uniformity

I love my leeks and grow several types that keep me supplied from October to April. For the most part they are all commercial cultivars and some, like the delicious and famous Scottish-bred Musselburgh, which was named after the area near Edinburgh where it was first sold in 1834, continue to be highly favoured among amateur growers. Incidentally, the Scottish leek is identified by its long, dark green leaf, known as a green flag, and its stubby stem which is called a short blanche. This differs from the 'London Leek', which has more evenly spaced leaves along its stem. The names of leek cultivars are both prosaic and descriptive. Most appearing in amateur gardening catalogues today have been bred in the last 200 years. Among my collection are Autumn Giant, which is now in its third incarnation as Autumn Giant III; Walton Mammoth from the USA; Lyon Prizetaker, an old French variety; and Colossal, an English variety bred in the 1980s. One of my favourites is an ex-commercial variety that stopped being sold 30 years ago: Walton Mammoth. The French are great leek-lovers too. Bleu de Solaize is a nineteenth-century classic, much like the Musselburgh in habit, but with blue leaves.

The leek is another vegetable that has been subject to modern breeding techniques to meet commercial needs imposed by supermarkets, allegedly to satisfy the demands of today's consumer. In 1993, a major breakthrough in leek breeding occurred when British plant researcher Brian Smith and his team discovered male sterility in leeks.[7] This enabled them to develop hybrid varieties with a resultant transformation in uniformity – an essential trait demanded by supermarkets who like their leeks to be delivered pre-packaged. Of more importance, I think, these new cultivars had improved resistance to some common pests and diseases. Today, 99 per cent of all leeks sold are hybrids. Modern leek production has increased enormously because they are easier to cultivate and their blandness appeals to the modern Western palate. Developing new varieties with disease resistance and greater yields is something we should never stop doing; growing and eating more leeks can only have positive health and nutritional benefits for us all, but let new cultivars be flavoursome too.

Smith and his team needed to grow thousands of traditional varieties to find the very few that had the male sterility trait. If it wasn't for the rich genetic diversity of traditional open-pollinated varieties, they would never have been able to succeed. Interestingly, their work at Horticulture Research International, now part of the School of Life Sciences at Warwick University, was supported by the multinational agrochemical company and plant breeder BASF who now hold one of the world's largest collections of leek varieties as part of their research gene bank. This means that they control the intellectual property of a crop which has all but replaced the genetically diverse and culturally rich traditional market. It's a familiar story. What the world is left to feed itself on are modern cultivars which are less genetically diverse than older varieties and thus less robust if some new bug, fungus, bacteria or weather event causes the crop to fail. If ever there was a reason to grow traditional and heritage varieties – and if you cannot grow them, to seek out and buy – then the leek is a prime candidate. Until you have eaten one you have no idea what you have been missing.

The number of leek cultivars available for people like me to grow has never been extensive. Even the wonderful Seed Savers Exchange in the USA only sells four. The 2015 EU Common Catalogue of Vegetable Varieties lists 338 varieties of leek with their provenance in countries all across Europe. Ninety-four of these are modern hybrids. One might take comfort from the fact that this list is long, but that doesn't mean all the varieties are in cultivation. On the contrary, only a handful are. The sad truth is that most are in the control of multinationals like BASF and Bayer, who most certainly do not want to share. Possession is nine-tenths of the law or, in the case of seed companies, 'finders keepers'. Embarking on a traditional leek-breeding programme for open-pollinated varieties, which really was the only way until 1993 when work on creating hybrids began, is a big call. Commercial companies won't do it because they lose control of the genetic material once it is in the public domain. Modern hybrid cultivars are legally protected by international intellectual property laws, which means they can only be grown with the permission of the seed producer under licence.

Maybe, though, I really don't need to worry about the future of this magnificent and ancient vegetable. Like its fellow alliums, the leek has been with us for as long as almost any other food; first growing wild and then, over several thousand years of domestication, becoming a bedrock of many food cultures. Maybe the story of leek breeding reminds us that there needs to be a place for both the most advanced and scientifically focused approaches as well as small breeders working on accessible, open-pollinated varieties that can be grown by the many rather than the few. The most obvious benefit of this is that we get to savour a wide variety of delicious and nutritious veggies and, alongside this earthly pleasure, comes greater food security. If we only grow one or very few varieties of any crop, including leeks, all it needs is a mutated pathogen or bug that takes a special liking to said cultivar for the world to go hungry.

A Never-Ending Search

I have been on a quest to find a genuine leek that was first bred in Wales – the John and Jane Doe of the vegetable world: missing, but hopefully not deceased. This has only got me as far as the 1939 Carter's Blue Book of Gardening seed catalogue where five varieties of leek are listed. In very small print at the bottom of the page is evidence that it did once exist: 'The Welsh Leek…A fine improved variety of the old WELSH' [their capitals] for the princely sum of 10d [4 new pence] an ounce [25g]'. Sadly, no longer available to buy anywhere. This, the most recent reference I have been able to find of a commercial Welsh leek, is proof enough that they were commonly available before World War II. Despite extensive research and endless enquires to leek growers across Wales there is no evidence to suggest a local heritage or heirloom leek has ever been bred. However, my search for this elusive vegetable continues. Maybe in some forgotten corner of a garden shed in deepest Wales a Carter's seed packet still survives…

Of Caulis,
Krambē and Braske

*A cabbage may be grown anywhere and
anyhow; that it will thrive on any soil, and
that the seed may be sown any day in the year.*

Sutton & Sons – *The Culture of Vegetables
and Flowers from Seeds and Roots* (1884)

School dinners. Not for the faint-hearted and for those
of us with long memories, a compulsory and miserable
gastronomic journey that's best forgotten. The school
where I spent my formative years believed in the benefits of
wholesome food, which meant slabs of brown bread that could
sink a battleship, quantities of sloppy stews of indefinable animal and…cabbage. Lots of it most days, come rain or shine.
Being a sad attention-seeker, I ate it in considerable quantities.
My nickname was 'dustbin' because my fellow students could
rely on me to eat sufficient cabbage from the bowl on the
table to convince our teachers that all of us had partaken of
the overcooked, soggy mush which was meant to be doing us
so much good. Thanks to those school dinners, I have spent
much of my life on a quest to find members of the fabulously
diverse and colourful cabbage family that are a culinary delight
– and with some considerable success. Generations of hungry

85

children have carried their prejudices against this vegetable with them, many to their graves, thanks to some truly criminal cooking. Fortunately, my experience of eating cabbage left no lasting scars. Patiently tended in the garden, then lovingly prepared in the kitchen, this, one of the world's most abused, misunderstood and unappreciated of all cultivated crops, has few peers. It was to take more than a decade after leaving school before I became aware that there really was a world of brassica bliss waiting to be discovered and a wealth of stories that underlined their place in our food culture.

Carl Linnaeus (1707–1778) classified all edible forms of the species *Brassica oleracea* into seven distinct groups. The ones from which we eat the balls of leaves that we call hearts are cabbage, and these are in the Capitata Group; non-hearting varieties like kale are the Acephala Group; the savoy cabbage Sabauda and kohlrabi, Gongylodes Group. The other three, of which we eat the flower buds, are broccoli, of the Italica Group; cauliflower, of the Botrytis Group; and Brussels sprouts, from the Gemmifera Group.

It's All in the Name

Etymology – the study of the history and origin of words – plays a vital role in understanding the journey of domestication as well as the use of language to describe vegetables. Etymologists have yet to find any mention of brassicas in the Bible: they are not included in descriptions of vegetables in Babylonian gardens or Assyrian feasts dating from 4,500 years ago, nor are they identifiable in Egyptian lists of edible crops. The Hindu Upanishads, written in Sanskrit between 2,500 and 2,800 years ago, doesn't mention brassicas either. The vegetables are absent from the Homeric poems, which are among the earliest works of Greek literature and contain about 50 botanical names. So, when were brassicas first domesticated and where?

The first Greek word that reliably describes leafy greens is *krambē*, which dates from about 2,600 years ago. *Kaulos*, meaning 'stem or stalk', first appears in Greek literature about 200 years later. The world had to wait until the comic Roman playwright Plautus (*c*.254–184 BCE) first used the word brassica to describe many types of cabbage-like vegetables; followed by Pliny the Elder (23–79 CE) a couple of hundred years later. The Romans also used the word *caulis*. Lorenzo Maggioni of Biodiversity International hypothesises that the possible link between the Greek and Roman languages is the Latin word *braskē*.[1] It was used by Hesychius, a Greek lexicographer who lived 1,600 years ago, to describe brassicas used by ancient Greek colonies in southern Italy. Maggioni considered it a reasonable etymological link, *krambē* to *braskē*, and then to the Latin *brassica*. Apparently, exclaiming in Greek 'by the krambē!' was meant to be funny, a comic turn of phrase that has not survived the test of time.

The upright stems of leafy kales were the most obvious traits of the first domestications, and it would seem that describing them in general terms as 'stems' became ubiquitous. 'Cole', which is derived from the Greek word *kaulos*, has remained a generic description of all types of brassica throughout the world and gives us the English name, kale. Today, the French call cabbages *chou*. In Irish the cabbage is *cal*, in Italian, *cavolo*, and in German, *Kohl*, with various local variants on the word cole in the Scandinavian languages.

From some of the earliest Greek literature we know that kale was a well-liked and popular vegetable, and it was boiled and eaten with olive oil. The first kales had either smooth or curly leaves. Of colour variations there is no evidence. The Greek cynic Diogenes (412–323 BCE), whose philosophy eschewed the trappings of fame and fortune for the simple life, apparently scolded a hedonistic friend's behaviour, saying: 'If you lived on caulis, you would not be obliged to flatter the powerful,' to which the young man is said to have replied, 'If you flatter the powerful you will not be obliged to live on caulis.'[2]

The wild form of brassica was determined by Linnaeus in his *Species Plantarum*, published in 1753,* to be native to the coastal regions of the British Isles. It was subsequently mapped as also native to some western parts of southern Scandinavia and the coastal regions of northern France and Spain. There was a presumption by some academics at the time that the Greeks took the word brassica from the Celtic word *bresic* meaning 'cabbage'. This is probably why Linnaeus believed the cultivated forms of brassicas were domesticated from wild types that he found growing on the white cliffs of Dover. Trade between the British Isles, the Phoenicians, the Greeks and Romans had been taking place long before the Romans invaded Britain. The exchange of plant material, accidental or otherwise, would have been common. More likely is that domesticated kale developed by Greek and

---

* Although no longer in print, *Species Plantarum* can be acquired from many sources as a facsimile. It was published in two volumes: the first in 1753 and the second in 1762/3. The work was the first to apply the binomial nomenclature system of naming plants in a consistent and systematic way. Before Linnaeus the naming of plants could be long-winded and inconsistent. Linnaeus introduced a two-part naming system: the first being a single-word genus name, as in this case *Brassica*, with a second single-word specific epithet. He used the word *oleracea* for the wild cabbage.

Roman farmers arrived with Phoenician traders who reached Britain by sea 2,500 years ago. The Celts would have taken their word for the vegetable, *bresic*, from the Latin *braskē*.

The wild inhabitants of Britain's coasts were also highly variable, some looking like kale with large leaves, others more like broccoli. Let us not forget that Dover was a major Roman garrison and the surrounding gardens and fields would have been used to grow the many varieties of brassica that the Romans brought with them at the time of their occupation. It was not a long journey for escapees from those plots to find a new home on the nearby white cliffs where they would revert to a wild state. Foraged wild cabbage from the area was much valued for its eating qualities, which is hardly surprising considering its likely parents were domesticated Roman introductions!

The Wild Children

All wild and domestic brassicas are highly interfertile, meaning they will cross and hybridise with each other for a pastime. Over the centuries, this has resulted in a vast number of local varieties. Anyone, including myself, attempting to grow brassicas for seed knows how important it is to keep the crop completely isolated to avoid accidental cross-pollination. Polymorphism in brassicas – the existence of two or more different forms of a species in a single population –is one of the most striking examples of crop variability caused through the action of human selection.[3] Research has now shown that domesticated brassicas which escape from a farmer's field and are left to their own devices go feral – becoming in effect a new native wild species. The British botanist James Syme observed in 1863 that 'Red cabbage of neglected gardens at the seaside pass back in a few generations to the condition of wild cabbage.'[4] What is now clear, thanks to gene mapping, is that the presumed 'wild parents' of today's brassicas are, in fact, the 'wild children' of earlier escaped and naturalised domesticated crops. Mapping of sites within the British Isles where there are populations of wild brassicas close to towns and villages shows they are the result of earlier cultivation from Roman and Saxon introductions.[5] Linnaeus got it wrong, therefore, and must take the rap for perpetuating the false hypothesis that the wild cabbage he found first on the cliffs above Dover was the parent of all the domesticated brassicas.

Domestication had, in fact, begun in the warm and sunny climes of the southern Mediterranean where kale thrived as a biennial– a plant that flowers in its second year. Through selection by farmers, responding to public taste as we shall see, it became a fast-growing annual. In northern regions where winters are long and temperatures not so clement, the biennial habit was exploited by German growers 500 years ago to give us the traditional hardy cabbages that we eat all year round. All types of brassicas flourish in all seasons and across hugely varying climatic conditions in seemingly endless incarnations. There is no other vegetable I grow that has evolved like brassicas through a mix of domestication, reversion to the wild and re-domestication.

Off One's Kale

Our love affair with eating leafy brassicas in all their forms goes back at least 4,500 years. Kale most closely resembles the wild parents that archaeological and historical evidence now suggests were first domesticated in their native home-lands of the Middle East and eastern Mediterranean. It is likely that, prior to domestication, the wild parents would have been foraged principally for their oily seeds and that early domestication of mustard – a very close relative – could have begun about 12,000 years ago.

As a kid, I remember the national collective view of kale as being fit only for cattle and if anyone was idiotic enough to want to eat the stuff, the experience would be a sorry one. Tough, bitter, unpalatable and not improved with much boiling. By the time I could afford to pay for my own dinner, Italian chefs had begun to test the conservative British palate with something called *cavolo nero* – literally translated as 'black cabbage' – a type of kale.

I was curious about this elegant Italian variety with its slender, ribbed, dark leaf. In the 1980s, cavolo nero was a horticultural rarity. It was really quite tasty, easy to grow, and withstood the worst that a British winter could throw at it. But was this the only kale worth eating? Italians will no doubt hate me for saying this, but of all the wonderful varieties of vegetable emanating from that country cavolo nero is by far the least interesting. I give it no space in my garden, preferring the many other delightful and tastier varieties I have discovered over the years. The Americans bred their own delicious types, which they called collards, from the English word *colewort* meaning 'cabbage'. A staple of Southern US cuisine, collard greens sprinkled with vinegar is my kind of food. One I grow regularly goes by the wonderful name of Georgia Southern Collard.

Travelling in the former Soviet Union, I soon became familiar with a number of different Russian variants. Unlike the narrow-leaved and almost tubular cavolo nero, Russian kale came in red and green, with large, serrated and frilly leaves. The Canadians had their own varieties too. Then, in the mid-1980s, I discovered asparagus kale. This awesome brassica, bred in Scotland towards the end of the nineteenth century, was so named because the flower spikes which emerge in spring can be eaten blanched, like asparagus. Delicious. But, for me, the real joy of this variety is its profusion of tender, pale green leaves that are at their best when plucked in late winter and stir-fried with garlic. I have been an advocate of asparagus kale ever since. It's not the only brilliant British kale. Another is Ragged Jack, so named for its deeply serrated, dark green leaves. Not grown commercially for more than a century, it survived in Tunley in Somerset where it was known as Tunley Greens. Another local variety is Black Jack from Tiverton in Devon, though back in the 1970s the grower was evidently keeping the crop to themselves because I never came across it when I lived in the neighbourhood.

Maybe because of a cultural distaste for kale in the U.K., in the Scottish vernacular 'to be off one's kale' is to be off one's food. Yet kale's mixed public image didn't stop it from being fundamental to Scottish cuisine. Curly varieties of kale are also known as 'Scotch kale' and for centuries there was hardly a meal eaten in Scotland that did not include a form of kale soup. The American botanist Edward Lewis Sturtevant (1842–1898), in his seminal work *Sturtevant's Edible Plants of the World*, completed in 1887, describes a traveller in Scotland called Ray who wrote in 1661 that 'people used much pottage made of coal-wort which they call keal'.[6] So ubiquitous was kale that not only was the generic name for soup *kail*, but also the pot in which it was cooked. Even Scotland's greatest poet Robert Burns (1759–1796) wrote of the vegetable's importance in assisting in the amorous intentions of young lovers in his poem *Halloween*. Now, there is an ever-increasing number of foodies who love their kale, some even seeing it as a 'superfood' and eating sprouting seeds or making lurid green smoothies from the leaves, which is apparently meant to do them no end of good. For some, their obsession for this once-despised vegetable extends to eating kale crisps with plenty of salt. I cannot think of another vegetable whose image has been so transformed in recent times.

Off with Their Heads

The Romans also thought highly of several different varieties of brassica, considering them another superfood, which thanks to generations of selecting and breeding had by this time spawned other forms, including the first identifiable cabbages. These originated, according to Roman legend, from the sweat of Jupiter; the result, apparently, of having to placate a couple of argumentative oracles.

The English word cabbage comes from the Latin *caput*, meaning 'head'. The first cabbage to be eaten in the spring from an autumn sowing can have wonderful pointy heads, which are often harvested as spring greens – those lovely bunches of delicious leaves that have been long been celebrated as harbingers of a new season of crops. These hardy types can survive the worst of winters and are generally fondly received. Varieties, some dating back to the late nineteenth century, have names as delectable as their flavour: Greyhound, Durham Early, April and Wakefield. Sadly now lost to cultivation, Wakefield was 'improved' in the USA during the latter part of the nineteenth century, to give us Early Jersey Wakefield and Charleston Wakefield, which are both still grown today. Nowadays, plant breeders have developed hybrid cultivars such as Caraflex and Advantage that grow very quickly, meaning 'spring greens' can be harvested all year round.

Selection by farmers in northern Europe during the Middle Ages led to hardier varieties. The first definitive description of cabbages was made by the French botanist Jean Ruel (1474–1537). In his book *De natura stirpium* (The Nature of Plants), written in 1536, he speaks of specimens that are 45cm (18in) in diameter and of a loose-headed type which he called *Romanos*, pointing to an Italian origin. Sturtevant suggested that because of the warmer temperatures in southern Europe, cabbages that had been developed from kale were loose-hearted. An example is the popular Savoy cabbage, which was domesticated between the eleventh and fourteenth centuries from a native wild parent. It was named after the House of Savoy, a mountainous region that covered parts of northwest Italy, France and Switzerland. There have been many wonderful descendants, including one of the most 'British' of all hardy cabbages, January King, which was first sold in 1867. I have grown this glorious variety every year for the best part of half a century and it's my favourite of all the loose-hearted cabbages. The Savoy cabbage, or *chou de Milan* as the French named it, was being widely cultivated during the seventeenth century in France, Holland and Italy. Wild escapee hybrids that populated further north and at greater elevations survived in much colder winters. A century earlier, thanks to the cabbage's polymorphism, new varieties of the more familiar red and white, solid-hearted cabbages made their way into our diet.

Cabbage, the key ingredient in coleslaw and sauerkraut, was a bedrock of northern European cuisine, thanks to the efforts of sixteenth-century Dutch and German growers. Sauerkraut's keeping qualities meant that pickled and fermented cabbage became a core part of peasant cuisine and was a vital addition to a ship's stores. Sauerkraut was extensively used by colonising and warring European navies from at least the fifteenth century for feeding mariners because it was nutritious and could keep for a year or more. *Sour Krout*, as Captain James Cook called it, was an indispensable food supply on his ship the *HMS Resolution* when he made his second navigation to explore new worlds between 1772 and 1775. He swore by its 'antiscorbutic' properties which warded off scurvy. Sauerkraut does indeed contain vitamin C, but whether this was the reason Cook was untroubled by scurvy is another matter. His sailors were able to eat fresh vegetables during regular landings and drank quantities of vinegar which also contains vitamin C. The prevention of this awful disease, which killed countless sailors during long journeys made around the world during the seventeenth and eighteenth centuries, came about as the result of the world's first-ever clinical trial undertaken by Dr James Lind and detailed in his *Treatise of the Scurvy* (published in 1753). Lind was able to prove that orange and lemon juice could cure scurvy and all that was needed to prevent the disease was a weekly dose of a fluid ounce of preserved lemon juice. He is also a distant ancestor of mine.

An Explosion of Brassicas

Until 1631 when John Winthrop Junior, son of the first governor of the Massachusetts Bay Colony, packed his bags in England to join his father, cabbage in all its forms was unknown in the United States. 'Cabedge' was included in a supply of seed he bought from a certain Robert Hill Grocer dwelling at the "three Angells in lumber streete'.[7] Perhaps seed merchants and pubs continued to be inextricably linked because in 1677 the seedsman William Lucas – whose address also appeared to be a pub, The Naked Boy near Strand Bridge, London – had several cabbage varieties for purchase and, a century later, Stephen Garaway of The Rose, near the Globe Tavern, Fleet Street, London, was also selling a choice selection of brassicas.

The explosion in plant breeding that took place across Europe and North America during the nineteenth century saw a blossoming in varieties of all brassicas. America's most famous seed company W. Atlee Burpee, after which many vegetable varieties are named, bred the first all-American cabbage. Called The Surehead, it was a tight-head variety which was first sold in 1877. It was a big hit, alleged to produce enormous heads that weighed up to 16kg (35lb). Within a decade the Burpee catalogue listed 32 cabbages of all types. The giant cabbage was here to stay. Today we have the red cabbage Drumhead, and I have grown a sizeable green heirloom, Paddy – no prizes for guessing where that one comes from – as well as an ex-commercial, green Drumhead type called Earliest that was popular in Britain in the 1930s and '40s. Seed catalogues of that time were awash with cultivars. Carter's Blue Book for 1939 lists 34, including many that had been in cultivation for at least the previous 200 years. Today's catalogues are rather different. Vilmorin Andrieux's Vertus is still available, along with many old English favourites, but more prevalent, however, are F1 hybrids, which promise higher yields, uniformity and greater disease resistance.

A Starchy Relation

Pliny the Elder (23–79 CE) in his encyclopedia *Naturalis Historia* (Natural History) described a particular type of brassica as '...[one] in which the stem is thin just above the roots but swells out in the region that bears the leaves, which are few and slender.' Based on his description, some academics have speculated that a type of kohlrabi was being grown in Roman times too. Others have suggested that Pliny was describing an early cauliflower. Also known as the turnip-rooted cabbage, the kohlrabi is, in fact, a form of stem cabbage in which the swelling we eat is a modification for the storage of starch. Other academics have suggested that a description in *Capitulare de villis*, an eighth-century document thought to have been written for King Charlemagne, of a type of kale with a turnip-like growth could have been kohlrabi. In his day it was being grown as a fodder crop.

Kohlrabi was also being eaten by us in the sixteenth century, as it too is described and illustrated by the Renaissance herbalist Leonhart Fuchs (1501–1566). Not remotely a fashionable vegetable, it would appear from the names of three varieties I grow myself – White Vienna, Green Vienna and Purple Vienna – that the Austrians at least liked and bred it. Personally, I love kohlrabi for its subtle flavour and tender flesh. Perhaps the most misunderstood of all brassicas, less is more when it comes to preparing it. Simply steamed or thinly sliced and eaten raw, it is a delight. Good in a stir-fry, too.

As we have seen, compared with many other vegetables in my garden, the humble cabbage and its ilk are the result of relatively recent domestication. In just a few centuries after the fall of the Roman Empire, skilled, curious and ambitious farmers living in southern Italy succeeded in breeding an entirely new group of brassicas which today are loved and loathed in equal measure around the world.

Fabulous Flowers

Selection pressures are external factors that affect a plant's ability to survive in a given environment. Negative pressures reduce the occurrence of a particular trait, whereas positive pressures increase the occurrence. Selection pressure can also be either density-dependent (where population size affects change) or density-independent (unaffected by population size). There are four main drivers of selection pressure: the availability of resources such as the presence of sufficient plant nutrients; the type of habitat and the ability to be effectively pollinated; environmental conditions, which include temperature, rainfall, competition from other plants and soil type; and, finally, biological factors, including pathogens and the action of herbivores, insects and other predators. It is thanks to the skills of curious farmers continuing to select for specific traits after the fall of Rome as well as the effects of selection pressure, principally the Mediterranean climate that enabled earlier onset of flowering, that we have broccoli and cauliflower. The word broccoli comes from the Italian word *broccolo*, meaning 'the flowering top of a cabbage' and the Latin *brachium* which translates as 'arm or branch'.

The first descriptions of the flowering types of cabbage do not make it clear if the Romans were describing broccoli or the cauliflower which, with its white or creamy curds, was a later creation. The Romans also had a great appreciation for the tasty and tender immature flowers of both the turnip, which they named *cymae* or *colliculi*, and kale. Gardeners often discard turnips when they 'bolt' and send up flower spikes. A big mistake because the immature flower heads are quite delicious in a variety of preparations. Similarly with kale, and my favourite, asparagus kale. Traits of early flower production are most likely to have evolved through accidental outcrossing between turnips and kale to give us today's foodie favourites, cime and broccoletti. They are close cousins of one of the delights of early spring: purple and white sprouting broccoli. These very hardy varieties are the closest relations to the early forms that were being enjoyed by the Romans 2,000 years ago. The large, green and occasionally purple heads of broccoli, usually wrapped like mummies in cellophane and piled high in the vegetable section of your local supermarket, are also called calabrese, which refers to their place of origin, Calabria in southern Italy.

One of the effects of selection pressure is the adaption of the leafy brassicas grown in southern Europe from biennial to annual traits. Long, hot Mediterranean summers would have encouraged many types of kale and cabbage to go to seed or 'bolt' in their first year – a problem most gardeners are familiar with. The farmer would then select from those 'bolters' that produced delicious, larger and more uniform flowering heads – what botanists call an inflorescence – and showed a stable annual habit. The renowned twelfth-century Arab agriculturalist Ibn al-'Awwām' used the Arabic word for cauliflower, *qunnabit*, to describe two kinds of flowering brassica that he called 'Syrian cole': one with a head which was closed, the flower buds *'all huddled together'* (that is, the cauliflower) and one with 'a head that splits in various branches' – broccoli. It could be argued, therefore, that it was in the eastern Mediterranean, including Syria, rather than southern Italy, that these plants were first grown. And, of course, the Roman Empire included the Levant, so there would have been a free flow of seeds around much of the Mediterranean. Renaissance botanists identified cauliflowers, giving them names including *Pompeiana* and *Cypria*, suggesting regional variations. In sixteenth-century England, cauliflowers and broccoli were referred to as *Cyprus coleworts*. Indeed, Cyprus was thought to be the source of the most highly prized seed.

Recent genetic analysis suggests cauliflowers were first bred in southern Italy as the result of crossing between a local broccoli and one from Sicily.[8] Cauliflower and broccoli, which come in a seemingly endless array of shapes, sizes and regional variations, would have been bred throughout the Mediterranean at the time of the Roman Empire. Although born in Seville, Ibn al-'Awwām probably called them 'Syrian coles' because he would have associated them with his native homeland. I also imagine that the Romans might have tried to introduce these crops, along with cabbage, in their more northern territories, but these early types probably did not adapt to the cooler conditions. Selection pressure takes time to work its magic. It was only the result of centuries of breeding hardier varieties that enabled flowering brassicas to travel north and establish themselves in all European cuisine from the time of the Renaissance.

Anyone who grows cauliflowers or broccoli will know that modern hybrid cultivars have a very annoying habit. They all mature at the same time. As I have discussed in other chapters, modern mass-production of vegetables demands that the farmer must harvest a crop all at once. Before mechanisation, cauliflowers and broccoli would have been harvested over many weeks, rather than just a few days, as every head matured at a slightly different rate. Much of the harvest would have been sold and consumed locally. I do grow the occasional F1 hybrid for an early season crop, but never more than four plants at a time. There is, after all, only so much cauliflower I can eat in a week! Much more rewarding has been seeking out traditional open-pollinated varieties, and it is not just to Italy that I have turned my attention when on the trail of great-tasting examples. My favourite cauliflower, English Winter/Late Queen, is hardy, delicious and huge and was first sold commercially in the U.K. in 1896. It is a true bragger's delight because it produces delicious heads that can feed a dozen hungry mouths. It also holds well – the curd doesn't open out into a seedhead quickly – which means it can be left in the ground for longer before being harvested. Most importantly, the heads develop over several weeks in late spring at the height of the 'Hungry Gap' when there is often not a lot of choice in the vegetable garden.

English-grown cauliflowers are considered by many gourmands to be the finest in the world. How could I disagree? When it comes to a visual feast as well as a culinary delight, however, it is to Italian heritage varieties that I turn. Above all others I value the traditional southern beauty, with a name that tells of its roots, Violetta di Sicilia. Like so many of these wonderful brassicas, the place of cultivation is invariably in the name. For example, the beautiful green cauliflower with its spiralling head, Romanesco, is from Rome, as is its white equivalent De Jesi; the creamy Verona Tradivo is from northern Italy; and from the Alto Adige region of the Dolomites comes Palla di Neve Adige.

A Tiny Cabbage to Love or Hate

You might think that the final brassica in this chapter doesn't belong among the flowering groups, but bear with me and all shall be revealed.

In the mid-1970s I ran a small market garden from the family farm near Tiverton in Devon. At that time the most basic of brassicas seemed to be the only thing visitors to the weekly market were prepared to buy: vibrant bright green spring cabbage; white, bland hearting cabbage; the occasional red cabbage for pickling in the autumn; the British stalwart, January King, in the winter and, of course, Brussels sprouts, which came in only one colour then – green. That was until I dared to attempt to broaden my customers' palates by offering red ones. Needless to say, they were viewed with great suspicion – I virtually had to give them away. I came to hate rising in the dark, trudging into the field to pick the dammed things on freezing winter mornings, and then being given much abuse by the public for not growing 'proper' sprouts. My life as a market gardener was short-lived and unprofitable, but today, trendy chefs and a curious public are warming to the twenty-first century sprout, as they are considered to be far superior to their more traditional green cousins.

As with so many vegetables, determining exactly when sprouts were first bred is still a matter of debate. They might have been known to the Romans, but their descriptions are so vague that they could equally have been describing sprouting broccoli. Yann Lovelock suggests that 'sprouts', as the English usually refer to them, could have been a development from a small-headed Milan cabbage, possibly in the sixteenth century.[9] Others have pointed to a possible start time for them in Belgium in the thirteenth century. I have my doubts because at that time the possible areas of origin, which were known as Middle and West Francia, were vassal fiefdoms of the King of France or the Holy Roman Empire, depending on who was Top Dog in the neighbourhood at the time. Brussels was an insignificant settlement on the banks of the river Senne and part of the Dutch-speaking Netherlands. Maybe a Flemish farmer from Brussels spotted a mutant cabbage on his plot and, as they say, the rest is history, but again, in the same way as the Dutch choose to believe that they bred an orange carrot to symbolise the House of Orange in the seventeenth century, Belgians claim sprouts.

There is some agreement that sprouts were being cultivated in Flanders and northern France in the sixteenth century.

Of Caulis, Krambē and Braske

Sprouts were grown in Louisiana by French settlers from the end of the seventeenth century. They are first described in Britain in a 'How to' gardening book, *Plain and Easy Introduction to Gardening*, written by Charles Marshall in 1796. The most popular way to eat them then was with butter after a good boiling 'in the Belgian mode'. Traditional nineteenth- and early-twentieth-century, British green varieties come with names that clearly place their origin: Severn Hills, Bedford Fillbasket and Evesham Special. The Europeans liked their sprouts small, whereas the Americans preferred them big and blousy. Although most breeding in the last 200 years has been in Europe, the Americans have developed their own large sprout cultivars from old European varieties. Today there are very few traditional open-pollinated varieties for sale in Europe or the USA; the lovely Long Island Improved being the only American variety I know of.

Brussels sprouts are, in fact, tiny cabbage buds that sprout from the stem of a type of loose-head cabbage with a small head of leaves that are the final 'coup de gras' from a harvest. As a kid I remember my mother telling me I should always pick sprouts from the bottom up, so as to encourage nice firm buds. Harvesting starts in late autumn or early winter. Purists like me would never dream of picking a sprout before it had been touched by the first frost of winter, which sweetens it. Come the end of winter, when everyone in the house is heartily sick of eating them, all that is left is the delicious head or 'sprout top', and when that has been cut a bald and battered stalk is what remains.

As with so many other crops, improvements in plant breeding and the creation of F1 hybrids have had a dramatic effect on the type of sprouts we now eat. Modern cultivars have greater disease resistance, especially to one of the deadliest for all brassicas, the fungal infection clubroot. However, one of the reasons I don't grow modern hybrids is because the sprouts tend to mature over a very short time. Farmers want their sprouts to be ready all at once, so they can harvest the entire stalk. A common sight around Christmas, when the British go mad for sprouts, is to see entire stems for sale, complete with a full complement of uniform sprouts. These modern cultivars are also milder and less bitter than many traditional varieties, which is why they are so popular. The modern palate prefers blandness to character. Unpleasant to some, it is the sulphurous smells associated with cooking sprouts which puts so many people off eating them. The problem for the poor vegetable is that it contains quantities of mustard oil compounds – a characteristic shared with its close relative the turnip – which are released when the plant cells are damaged, usually as a result of much boiling. Nature evidently never intended us to eat sprouts, which employed these noxious compounds as a defence mechanism against predating caterpillars and other insects, as well as browsing herbivores. Breeding out those sulphurous traits in Brussels sprouts in response to a modern desire for our greens to be bland comes at a price. Those strong tastes, which are also present in many older varieties of broccoli and kale, are markers of high levels of antioxidants and an indication of nutritious plants.

A New Family Member

In the last few years, and in response to the vegetable's declining popularity, breeders have crossed kale with sprouts and created an entirely new brassica, the flower sprout. This is marketed as the kalette, thus detaching it from the sprout and its negative image. So named because the loose heads look like miniature florets, these sweet and pretty plants are proving something of a hit with the consumer. Even I am tempted to grow some in the hope that I will no longer be the only person in my family prepared to eat the dreaded Brussels sprout. Arguably, this humble vegetable is the first new vegetable to be created in the twenty-first century.

Thanks to 2,500 years of careful selection first by farmers and later plant breeders, the world has a seemingly endless palette of brassicas to love and hate in equal measure. They all come with powerful cultural, linguistic and culinary identities that are the bedrock of all our food cultures. Modern plant breeding has succeeded in creating cultivars that are reliable and profitable for large agri-businesses. Some are delicious, but the traits of uniformity bred into them are not useful to the ordinary gardener. Their ever-increasing presence in seed catalogues also presents an existential threat to those heavenly traditional varieties that are front and centre of my own identity and to the diversity of our food heritage. The genetically diverse nature and evolution of local types of brassicas that can flourish in a wide range of climates should be encouraged, not just because they are an important part of our sense of identity, but also because they provide greater food resilience as our climate changes.

It's a mystery to me why anyone would not love cabbages and cauliflowers, kale and calabrese – vegetables that nourish us and stimulate our taste buds when there is little to harvest at the end of a long, cold winter; that give us joy as harbingers of spring and remain intrinsic to the festival of Christmas. The command to 'eat your greens' remains as important as ever for our own health and well-being. Rest assured, I shall continue to live up to my nickname, Dustbin, unrepentant.

An Aspiring Spear

Nature gives us the key to every secret that concerns our happiness, and in respect of asparagus cultivation she is liberal in her teaching

Sutton & Son – *The Culture of Vegetables and Fruit from Seeds and Roots* (1884)

T
he dawn chorus heralds the arrival of the first spears, an eagerly awaited culinary delight for asparagus lovers like me. 'All good things come to those who wait', a quote by the Victorian poet Lady Mary Montgomerie Currie (1843–1905), is certainly true for those of us who love to rush to the kitchen with an armful of freshly cut asparagus in late spring and gorge ourselves on this magnificent vegetable. By the middle of April my first morning stop in the vegetable plot is at the asparagus bed. Can I see the early signs of pleasures to come – pale pinkish-green little noses pushing their way out of the soil? Must I make do with just a few spears until the crop is growing full tilt in May? Can I bear to share a surplus or shall I make more soup to freeze and enjoy that asparagus taste out of season, just like my distant ancestors, the Romans?

I have never found asparagus an easy plant to establish in my vegetable plot, although to read gardening books on the subject it should be. A member of the lily family, *Asparagus officinalis* is believed to be another native of the Fertile Crescent, where it was depicted in Egyptian hieroglyphs from the Memphis dynasties of the third millennium BCE. It was grown in Asia Minor and Syria more than 2,000 years ago and was much loved by the Greeks and Romans. The Greeks described asparagus as the fruit of a ram's horn thrust into the ground. Centuries later this myth was employed as an insult – thumbs and fingers outstretched on either side of the temples signifying the horn-wearing cuckold. The Roman elder Cato wrote instructions as to its cultivation in 200 BCE and Pliny the Elder, 200 years later, described three spears together weighing the equivalent of 325g (11½ oz). So, just as today, even in Roman times size mattered.[1]

Asparagus gets its name from the Greeks, who took the Persian word for 'shoot' or 'sprout', *asparag*, to give us aspharagos, which was then Latinised to *sparagus*. The medieval English word is a bastardisation of this, giving us sperage and later sparrow grass. Although it was the Mediterranean civilisations who principally embraced asparagus, it is also considered to be native to northern Europe and numerous other edible species of this genus can be found across the world. Sutton & Sons, one of Britain's most important plant breeders, had this to say about asparagus in 1884: 'It is so abundant on the sandy steppes of Southern Russia and Poland as to kill out the grasses, but it takes their place in respect of utility, the horse and the cattle eat it as daily food and enjoy life and prosper...'[2] Asparagus was growing wild in Britain 2,000 years ago when the locals would have used it as a medicinal herb and invading Romans would happily forage for it as a delicacy. It is thought by some historians that the Romans introduced asparagus to Britain as a cultivated crop, but when they left in 400 CE it would appear they took their love of eating it with them. There is no evidence that Britons ever foraged for asparagus as a vegetable, either before or after the Roman occupation.

Mad About the Spear

To say that the Romans were fanatical about asparagus is something of an understatement. Today we think of it as a cultured food, expensive and only for aficionados, with a traditional short season in the northern hemisphere between April and June.* But in Rome, asparagus was a culinary obsession too. Not only was it consumed fresh in vast quantities, but it was also dried to be later reconstituted for the dinner table. The Romans even froze their surplus. In the first century CE, charioteers and runners sped north from Rome to the southern Alps where the crop was kept under ice and snow to be eaten at the Feasts of Epicurus in January and February, several months before the new crop could be harvested. The emperor Augustus had an asparagus fleet of fast ships specially built to transport freshly cut spears around the Empire. He was a bit of a wag too, being credited with the command '*Citius quam asparagi coquentur*' (Do it quicker than you can cook asparagus). Some say that Julius Caesar ate his asparagus with melted butter. A recipe for how to cook the vegetable can be found within the pages of the world's oldest surviving collection of recipes, *De re coquinaria* (The Art of Cooking), thought by some historians to be the work of gourmet Marcus Gavius Apicius in the first century CE.

---

* Today, thanks to modern plant breeding and the method of growing under cover, asparagus is harvested in the U.K. between February and October.

Asparagus was, and still is, prized for its medicinal properties as a diuretic. The plant is rich in vitamin K which can help conditions such as varicose veins. In Rome, a syrup of young spears and extracts of the root was prescribed by herbalists as a sedative for those with heart disease. Even today some asparagus lovers swear by it as an aphrodisiac, a trait the phallic spear has been blessed with for at least the last 3,000 years. Its therapeutic qualities were legion. The Romans believed that putting the root on an aching tooth would ensure that its extraction would be painless. Bees would not sting anyone anointed with a tincture of asparagus and oil. The Romans also cultivated the crop in trenches – a method that dominated its cultivation until the nineteenth century – though why exactly fails me. Asparagus was to them what ginseng was and continues to be to the Chinese?

A Fabulous Family

Several wild species of asparagus are native to Europe and north Africa, and many form an important part of local cuisine. A sub-species found on the Atlantic shores of Spain and across the Mediterranean is seaside asparagus, *A. maritimus*, which was also foraged by the Greeks and Romans. Because it crosses with domesticated species, it is an important constituent in modern asparagus breeding. Another wild sub-species, *A. acutifolis*, was also cultivated by the ancient Greeks and is native to southern Europe and parts of north Africa. It has uniquely tender spears which become wonderfully aromatic when cooked. This species has also been foraged in its native lands for at least the last three millennia and continues to be sought after by foodies today.

Other wild species native to southern Europe include spiky asparagus, *A. horridus*, which is valued for its long and pointed shoots – nothing horrid about that! Then there is slender-leaved asparagus, *A. tenuifolius*, which can be found from the Mediterranean to the Ukraine. Wild white asparagus, *A. albus*, is native to the high plateaux of Algeria's Atlas Mountains and can be bought in the markets of Algiers in early summer. It is also found in Madeira where it too is harvested from the wild and known as garden hedge asparagus. *A. aphyllus* is indigenous to Greece and the young shoots are most frequently eaten during Lent.

Named *A. abyssinicus* (but now considered by some botanists as synonymous with another African wild species *A. flagellaris*) Ethiopian asparagus is native to countries surrounding the Red Sea and is much appreciated for its large root, which is enjoyed fried. The roots of *A. pauli-guilelmi*, native to parts of West Africa (also considered synonymous with *A. flagellaris*), are a boiled delicacy and this species improves with cultivation, suggesting a potential for further domestication. Beyond the Fertile Crescent there is a Far Eastern native, *A. cochinchinensis*, or shiny asparagus, which is enjoyed for its roots too; in Japan these are eaten candied. And the fruits of *A. racemosus*, more commonly known as rooted asparagus, are eaten by the Golo people from Darfur, Sudan.

With the fall of the Roman Empire, asparagus went out of fashion in Europe. It was thanks to invading Arabs, who reintroduced varieties from the Middle East to Spain in the eighth century, that this most noble of vegetables returned to discerning tables. It was not until the sixteenth century, however, that the European aristocracy went wild for this springtime delight and farmers began to develop new

110

varieties. Purple Dutch, as the name suggests, was bred in Holland and Germany, where it was known as the *Königs-gemüse* (king's vegetable). Its development fired the starting gun for a century of breeding, giving us an enduring type, the white-tipped German White asparagus.

The practice of growing asparagus in the absence of natural light by earthing up the spears to 'blanche' them has been favoured by gourmands around the world for the last 500 years. Not to my taste, but considered by those who eat blanched asparagus as being the best for flavour. The French nobility got a taste for the vegetable big time in the seventeenth century when Louis XIV had a greenhouse specially built so as to enjoy an early crop – though whether blanched or not is unknown. At the same time Samuel Pepys wrote of 'A hundred of sparrow grass' being sold in London's Fenchurch Street. This was definitely not blanched!

A Growing Craze

It was the French Huguenots fleeing religious persecution in 1685 who are believed to have started the market garden in England. This small-scale intensive system of horticulture was developed on a vast scale over the next 200 years. Asparagus did not have to travel far to reach the affluent diners of London and the South East in the eighteenth century – the light and sandy soils on the south side of the river Thames were much favoured for its cultivation, which continues to this day. Although large acreage was set aside for growing asparagus in Kent and Essex, the finest and earliest spears were grown in London's Battersea Gardens where, by the end of the eighteenth century, some 260 acres (105 hectares) were under cultivation. It was only in 1848, with the laying out of Battersea Park, that asparagus cultivation came to an end in the heart of London. Mortlake and Wandsworth to the west of the city were also great centres of asparagus cultivation, as was Deptford to the east.[3] Sadly no more.

Asparagus is first mentioned under cultivation in the USA by Adriaen van der Donck in his *Description of the New Netherlands*, published in 1655, where he lists asparagus growing in the herb garden. As plant breeders spread across the colonised world, it is hardly surprising that, during the seventeenth century, settlers in the New World were trading plants back and forth across the Atlantic. America was fertile territory for European plant hunters and, equally, with its varied but often similar climate to that of mainland Europe, welcomed garden crops introduced by immigrants. Native Americans quickly adopted asparagus for its medicinal qualities rather than as a food crop.

Asparagus only became a commercial crop in the USA in the late eighteenth century when another Dutch colonist, Diederick Leertouwer, first planted it in the town of West Brookfield, New England, in 1784. Local resident Ruth Lyon has done much detective work on the town's most famous horticultural resident.[4] It would see that Diederrick grew white asparagus, which was the preferred method of cultivation in Europe at the time. Ruth thinks it likely that the Dutchman imported seed along with other Dutch produce through one of his many businesses on the wharf in Boston, some 70 miles away.

The traditional way to propagate asparagus plants was from seed. Spears develop as shoots from a clump of roots known as a crown; the best specimens are selected by thinning or transplanting. When starting a new asparagus bed, I used to like to grow from seed because it is a lot cheaper than buying year-old crowns. It is also safer to grow from seed, to avoid importing possible diseases that could be lurking in the crowns. The three-year wait for a harvest with this traditional method of cultivation, however, has something of the Tantric about it and selecting the best seedlings to grow is a serious faff. It can also have malodorous consequences as traditional varieties have a most unpleasant trait, producing ammonia as they pass through the gut, which leaves the memory of spears enjoyed in one's pee. One of the great successes of modern plant breeding has been to create varieties that not only outperform traditional open-pollinated ones but are also, as we shall see later, less smelly. Needless to say, my asparagus bed is today entirely populated by bought-in modern crowns.

Breeding Gets Serious

Towards the end of the nineteenth century, a new 'English' variety of asparagus came onto the market called Connover's Colossal. Like so many other vegetables in this book, asparagus breeders and growers liked to appeal to a consumer's national identity when marketing their goods. This home-grown variety was actually bred in the USA and, although there are many far finer types today, it remains popular with gardeners 150 years later. In 1865, American seed merchants were also selling a variety called Barr's Mammoth. It is still on the USDA (U.S. Department of Agriculture) list of available cultivars, although now only grown by a few enthusiasts. Another American stalwart, which I used to grow, is Martha Washington, much appreciated by American growers a hundred years ago. This was named after America's first First Lady, who had a great interest in the cultivation of crops and would certainly have been curious about asparagus, but sadly not the one named after her – Martha had already been pushing up the daisies for over a century before this variety was developed.

The USDA played a key role in the breeding of new varieties of all sorts of edible crops, and in 1919 Dr J.B. Norton released the first of the new strain of asparagus known as the 'Washington' varieties. He called this asparagus Mary Washington, though why this name was chosen is a mystery to me. Martha Washington was also part of this 'family' of varieties. The 'Washington' strain was widely used in the US to develop new varieties better adapted to local conditions. Mary Washington became a key source for many modern hybrids and continues to be grown a century after its first introduction.

There has been much research over the years into the evolution and development of domesticated asparagus and its wild relatives. Although many wild and cultivated species cross easily with each other, both deliberate and accidental hybridisation have not resulted in greater diversity. The traditional method of breeding from seed, which had been employed for millennia, was creating an existential problem. By the middle of the twentieth century one of the big issues for the asparagus industry was that the quality of new varieties was, in fact, deteriorating as a result of the lack of genetic diversity in the parents. Size and quality were on a downward trajectory. Asparagus was in trouble.

Asparagus was in trouble because of its sexual habits. It is dioecious, which means that the plants are either male or female. A traditional asparagus bed would normally contain a mix of lads and lasses. This becomes evident when observing a traditional variety such as Connover's Colossal as it grows through the summer. As the days get longer, mature female stems will be festooned with bright red berries whereas the males are bare. Unless the grower has been diligent in only selecting one sex from the original sowing, they will find large numbers of self-seeded plants growing the next year – as I know to my own cost, having spent many a long hour weeding them out! The conventional approach to breeding asparagus was to select the best young plants, known as crowns, from year-old seedlings. However, the progeny are always hugely variable because they are the result of fertilisation from multiple fathers. No two seedlings are the same. Breeders had presumed there was no other way to select progeny that could be relied upon to produce a new generation of consistently better crowns, and it was all a bit hit and miss. Every time a new and desirable strain was grown its progeny were much like those that had come before. Something needed to change.

Celebrated British breeder A.W. Kidner, who is best known for developing the Kidner's Pedigree strain which won an RHS Award of Merit in 1953, took a very different approach to developing his famous prize winner and this was to completely transform how asparagus was bred, so reversing its downward spiral. He pointed out that a single spear weighing 100g (3½ oz) would be considered rather fine today. That is the same weight as a spear described by Pliny over 2,000 years ago. So, despite two thousand years of selecting crowns from, as Kidner described it, 'uncontrolled promiscuous fertilisation', the reality was that asparagus had not been materially improved with tastier spears, better disease resistance and heavier crops since Roman times. Kidner believed that this was further evidenced when comparing varieties grown all over the world in a diverse range of climatic and soil conditions. Essentially, they all behaved and tasted the same. As Kidner puts it: 'asparagus grows steadily on, apparently unchanged. Buds grown in sunny California under irrigation can hardly be distinguished from buds grown in the sandy soils of fog-bound England.'[5]

With concerns that asparagus breeding had found itself in a genetic cul-de-sac, Kidner maintained that to improve strains of asparagus a new approach was required. He needed to apply the same methods as a cattle breeder does to improving the 'line' of the herd: namely a systematic, controlled process of selective pedigree breeding. As with the cattle breeder, he sought to concentrate desirable characteristics in his new strain of asparagus through both parents of each generation; all parents exhibiting the same characteristics in a continuous line of controlled breeding.

It took eight generations of selective breeding to arrive at Kidner's Pedigree. His approach to developing greatly improved and consistent new varieties still employed a process of selection using open-pollinated methods. However, a greater understanding of plant genetics was to result in a fundamental change in asparagus breeding from then on.

A Thoroughly Modern Spear

One needs to return to the complex sexual characteristics of asparagus to understand how asparagus breeding has moved on since Kidner. Although asparagus is dioicous, Mother Nature can play fast and loose with an individual asparagus plant's gender. As well as being either female or male, some are bisexual or hermaphroditic. Others can be andromonoecious, which means they have both hermaphroditic and male flowers on the same plant. This has been an important factor in the development of all-male hybrid cultivars over the last 50 years or so.

Since asparagus can take up to five years to grow sufficiently to produce a decent crop, breeding new varieties is a slow job. Hybridising asparagus can take decades. Breeders cross andromonoecious plants with supermales – plants that have an identical pair of 'Y' chromosomes which determines their gender – and then carefully select their hybrid progeny over many years. When back-crossed* with the right kind of female, only seed that is going to grow as a male plant is produced. One of the ways breeders create sufficient numbers of hybrids to produce male-only seed is to take the parents of resulting superior lines and propagate them asexually using in-vitro culture in the lab. In this way they are able to create large numbers of clonal hybrids from which seed can be harvested. Before Kidner transformed the way breeders approached improving asparagus, uniformity and consistency in the asparagus bed was but a dream. Today, with the revolution in our understanding of plant genetics enabling us to control the sex of the asparagus plant, it is now a reality.

If you are a breeder who is investing a fortune in creating the next generation of high-yielding, long-living and tasty asparagus, you will want to hang on to the intellectual property of the cultivar. In the case of all-male hybrids like asparagus, it is impossible for the farmer or a rival breeder to save their own seed as none is produced. The other reason to grow only male asparagus plants for the table is that they tend to produce heavier crops, be fatter and live longer than female plants. The reason female plants are considered

---

* Back-crossing involves pollinating the hybrid with either the parent or a plant that is genetically very similar and stable and which has more desired traits, such as better size, taste and disease resistance.

'inferior' is because they expend much energy in making seed and thus their spears are generally thinner and fewer in number. Fat spears are considered to have softer, more succulent interiors, which some consumers prefer. Modern cultivars are, for the most part, all-male hybrids, initially bred from an eighteenth-century Dutch variety. However, any visitor to a food stall in the Far East who orders stir-fried asparagus will be given a plate of delicious, thin, traditionally bred spears, generally of both sexes. It's all a matter of taste.

Big Business at a Price

More asparagus is consumed in China and the Far East than anywhere else in the world. The USDA has calculated that the Chinese grow in excess of 250,000 tons a year, with the USA a long way behind, growing just 34,500 tons. It is the Peruvians who grow the most (390,000 tons in 2017) – encouraged by the USA who in the 1990s wanted Peruvian farmers to stop growing coca to produce cocaine and to make their money with asparagus instead. This has been seen as a great success story, with asparagus becoming a vital Peruvian export worldwide. However, in the major asparagus-growing region of the Ica Valley aquifers are literally being sucked dry by the vast amount of production. There may be much-needed full employment in this booming agricultural sector, but the ecological and economical challenges facing farmers and consumers add yet more layers of complexity to the question of how the world should feed itself sustainably. The price we pay to satisfy the desire to eat our favourite foods at all times of the year – just like the Romans – is becoming too high to be environmentally sustainable.[6] For me at least, the great joy of asparagus is its seasonality and for this reason I eschew imports.

As is the case with so many modern cultivars, the result of breeders focusing on a small number of traits means their genetic diversity becomes ever narrower. The latest asparagus cultivars are bred for herbicide and disease resistance as well as drought tolerance. Yet pests and diseases naturally evolve and mutate as part of the evolutionary process that continues in all living things. The trouble with pathogens – bacteria and viruses – is that they can, with lightning speed, mutate and develop immunity to specific herbicides and pesticides. For a plant to combat bugs and diseases it must have the genetic resources to be able to adapt to an ever-changing environment. The wider its range of inherited traits, the better. In the case of modern asparagus, which has been bred primarily to maximise yield over a longer season, its inherent weakness is an inability to survive in a changing world because of its narrow genome. To that end, plant geneticists seek out landrace and wild relatives to cross with modern cultivars in order to broaden their genetic diversity. To understand the qualities of these ancient relatives it is essential to devise an effective strategy for modern plant breeding. Deciphering the genome of traditional varieties to see who their wild parents were is the starting point.

Much natural hybridising has happened between cultivated asparagus and its wild relatives, resulting in exciting native landraces. This is especially the case with purple cultivars that were first bred in Spain and Italy during the last century, and in New Zealand for the last 60 years.[7] These include the Italian landrace, Violetto d'Albenga, and the Spanish landrace, Morado de Huétor. Researchers have determined from genetic analysis that both are the result of natural hybridisation in the wild between cultivated and seaside asparagus. Now plant breeders can use both these old landraces to increase the gene pool of a new generation of all-male cultivars. For example, if you find an asparagus called Purple Passion on your plate, it is the result of breeders working with Violetto d'Albenga. I would be surprised – subject of course to the chef – if you didn't think it delicious. Purple asparagus is not generally as heavy cropping as its green brothers, but it can be harvested later, in my garden at least!

The enduring strength and depth of our cultural relationship with the geography of our food is now also a political and commercial matter to be celebrated, protected and promoted. Huétor Tájar is a region in southern Spain near Granada that's famous for its asparagus. Crops grown here have special status, Protected Geographical Indication (PGI), granted by the EU, which confirms their unique place in Spanish cuisine. Asparagus from this region is not only good to eat, but also of much interest to plant breeders because it shows greater genetic variability and high levels of flavonoids – plant chemicals that are responsible for colour as well as being powerful antioxidants. Nothing has changed over the centuries when it comes to promoting asparagus's medicinal qualities and benefits for a longer and healthier life. The cultivar Triguero de Huétor, rich in those flavonoids, thought by some health professionals to have anti-inflammatory and immune system benefits, is also rich in saponin. This compound gets its name from the frothy excretions of soapwort, *Saponaria officinalis*, the root of which was used to make soap. It is an antioxidant but with a bitter taste that many plant species use as a defence mechanism against browsing herbivores – which includes us humans. Saponin also has antimicrobial and antifungal properties, something that could have health benefits for those of us who love our asparagus. Perhaps the most useful aspect of saponin is its ability to reduce the amount of ammonia excreted by cattle. For this reason, it is a common supplement in animal feed. Saponin also helps to reduce the 'asparagus' smell in our own urine, which adds to its commercial appeal.

Triguero de Huétor reminds me so powerfully of just how deep our relationship with the food we grow and eat really is. So apart from its unique health properties and its place of origin, what exactly does this modern Spanish spear look like? Certainly not obese; on the contrary, it is slender – dare I say, skinny? Size may have mattered for the Romans and asparagus growers over the millennia, but it is evidently not a must-have for the Spanish consumer.

Evesham in Worcestershire, in the U.K., is the largest centre for asparagus cultivation in Northern Europe and farmers have been growing it there since the middle of the eighteenth century. Despite the fact that a certain Anthony New of Evesham was awarded a medal in 1830 by the Horticultural Society of London for his asparagus display, there is no variety of asparagus directly associated with the town. This is, sadly, yet another example of how detached us Brits have become from our local and regional food culture, especially when it comes to vegetables. Evesham does, however, host the annual British Asparagus Festival,[8] where the best crops are auctioned off at good prices and the spear is celebrated with song, dance and much feasting.

The Evesham festival is, however, nothing compared with the great Spargelfests: asparagus festivals held across Germany during springtime in celebration of the white crop. The most famous event is at Schwetzingen, a town near Heidelberg close to the Rhine. Proclaimed as the world capital of asparagus, the festival is a homage to the blanched spear. Yet it is the green varieties that dominate the world market and, to my taste, are infinitely preferable to the pale, fat and rather bland white spears so treasured by many.

I cannot imagine my garden without an asparagus bed. Plants can grow very happily for decades in the same place, and they are the loveliest of all perennial vegetables. They can only be harvested and relished for a few glorious weeks, after which the plants are left to grow on for the following year. Their tall stems, waving in a summer breeze like forests of feathers, always bring a smile to my face. Then on Christmas Eve I ritually cut down the now-dead stalks and, like generations of growers before me, set fire to them on the bare soil. There is a good reason for burning them, because the scourge of asparagus growers, the colourful asparagus beetle, overwinters in the stalks. Any that I might harbour come to a fiery end during this season of goodwill.

Tending my asparagus bed requires diligent weeding throughout the year – they hate to compete with other plants. They are greedy too, requiring spring mulching with nutritious compost and the occasional application of soot from a wood-burning stove to add nitrogen to the soil. The rewards are always worth the effort. Depending on the weather, they start to appear in late March or early April. Patience is rewarded when I take my knife to the first spears to reach about 17cm (7in) in length. I can enjoy the finest harbinger of a new season just as Julius Caesar himself did, dipped in melted butter and consumed with relish.

For the Love of a Leaf

*'The lettuce is the king of the salads in the sense
that it may be eaten with relish and advantage
without any aid from the cook'*

Sutton & Son – *The Culture of Vegetables
and Fruit from Seeds and Roots* (1884)

Winter is the time to harvest some of the tastiest and loveliest crops. I love what my vegetable plot has to offer then, probably more than at any other time of the year. The excesses of Christmas and New Year are yet to become a distant memory and it may be miserable outside, but my forays into the polytunnel always lift the spirit. The temperature is chill and, like a damp towel, the fog rises from the Severn Estuary as I expectantly lift an upturned flowerpot on one of the beds. Revealed is a plump fist of tightly packed leaves, white with the palest green glow, as if lit from within – my first blanched Belgian chicory of the winter.

This glorious plant is not the only salad vegetable I will be harvesting. Nearby is an orderly row of grapefruit-sized, ruby-red balls of Italian radicchio alongside dinner-plate French escarole, some wearing white plastic hats of a similar size which hide a cluster of succulent pale green and tender leaves. Outside, sheltering under a long polythene cloche, is a row of English lettuce plants forming a six-inch-high hedge of deep green leaves tinged with pink and yellow veins. Tonight, I shall be eating a truly European salad. Nutritious, healthy and unimpeachably delicious.

I've got a bit of a thing about lettuce and its cousin, edible chicory, both of which include a wonderfully delicious and diverse number of varieties that populate my vegetable garden throughout the year. Today the whole world eats lettuce. This was not the case until a couple of hundred years ago, when plant breeders got their hands on the very few varieties that had existed up until then and embarked on a frenzy of genetic creativity.

Chicory and lettuce are members of the daisy family Asteraceae, the second largest group of flowering plants in the world. Lettuce is part of the genus *Lactuca*, a word that refers to the milky-white sap the plant exudes when cut. It is a mild narcotic that acts as a soporific and muscle relaxant, which is one of the many reasons lettuce has been intimately entwined with our social and cultural history from the days of its first domestication.

Beginnings

There is still much academic debate as to the identity of the cultivated lettuce's wild parents, probably because there are about a hundred wild species found around the world. The Fertile Crescent was thought to be home to the greatest concentration and *Lactuca serriola* was believed to be the first parent. But a number of other species underpinned the domestication achieved by Neolithic farmers about 8,000 years ago. There are also compelling arguments to suggest that lettuce was first domesticated in Kurdistan and Mesopotamia long before it made its first documented appearance in Egypt 4,500 years ago.[1]

Cultivated lettuce (*Lactuca sativa*) are classified into seven groups: butterhead, crisphead, cutting, Latin, stalk (aka asparagus lettuce or celtuce), oilseed and the first domesticated loose-hearting type, cos. Oilseed lettuce was widely grown in Egypt for the oil derived from its seeds, though it does not produce anything like as much oil as other crops the Egyptians would have been growing at the time, such as mustard. To me, it is a bit of a mystery why oil was extracted from lettuce, yet oilseed lettuce continues to be cultivated to this today. The other lettuce Egyptians loved, the cos-type, produces an abundance of tasty leaves. Its image first appeared in tombs over 4,500 years ago – what looks like a cos lettuce adorns a wall bas-relief in the tomb of the sacrificial chapel of Choe at Abydus from around 1800 BCE.[2]

The Egyptians ate the leaves and used the plant in sacrificial rituals associated with Min, the god of fertility and procreation. It was considered an aphrodisiac, probably due to its abundance of white, milky latex. Even today, Egyptians believe that the consumption of lettuce assures plenty of children. They eat a variety called Balady – meaning 'local' in Arabic – which looks much the same as the image of the loose-leaf, pointed cos lettuce at Abydus.

It is thought that the Egyptians introduced the cos lettuce to the Jews, who used its leaves among the bitter herbs eaten at the Feast of Passover. Wild chicory, which has a habit very similar to that of the cos lettuce and is bitter, was also being domesticated in the Fertile Crescent at the same time. It is therefore likely that the mixed leaves at Passover contained several members of the Asteraceae family. The cult of Ishtar and Tammuz, deities of the Sumerians of Mesopotamia and synonymous with the Greek gods Venus and Adonis, indicates that ancient civilisations venerated the lettuce too. When Tammuz died, he was allegedly laid on a bed of lettuce. To commemorate his death, lettuce was grown in pots, which would then be carried in a procession as symbols of the transitory nature of life. The Greeks named these lettuce-filled pots 'The Gardens of Adonis'.[3]

When the Greeks were cultivating lettuce 2,500 years ago, they considered it an anti-aphrodisiac and soporific.[4] There are a number of references in both Greek and Roman literature to the medicinal qualities of lettuce and it was also considered good for the digestion. Lettuce was beneficial, supposedly, in maintaining bowel health too. The Romans ate lettuce at the end of a feast in order to stimulate sleep, although one must wonder whether eating it after a large meal made sleep come any sooner. Later, they started to eat lettuce with radishes at the beginning of a meal to stimulate the appetite! What a plant – both stimulant and soporific!

The Chinese were eating lettuce at the same time as the Greeks, but favoured the stem rather than the leaves. A tall, fleshy lettuce with a thick stalk is native to Tibet and would have been crossed with cultivated lettuce from the West, enabling farmers to focus on selecting for flavoursome fleshy stems. Known as celtuce, the leaves were also eaten, cooked like spinach. It was also cultivated in Russia and India and was brought to Europe in the nineteenth century when lettuce breeding was really taking off, but failed to appeal to the local palate. Celtuce is now quite a trendy plant and can be found in some fashionable gardens across the northern hemisphere, but mostly on allotments cultivated by the Chinese diaspora. It found its way into American cultivation in the late 1930s, but it is really only grown by amateur enthusiasts. I haven't grown it yet, although I think it might be fun to try.

Changing Shape

My favourite lettuce gets its name from the Greek island of Kos where this cylindrical type has been grown for at least the last 3,000 years. The Romans called it the Cappadocian lettuce; according to Herodotus (*c*.484–*c*.425 BCE), Cappadocia was the name used by the Persians to describe the lands of Greece and the eastern Mediterranean. A visitor to this part of the world cannot fail to notice the mountains and hillsides covered with ancient, abandoned terraces: the valleys speckled with carpets of small fields. These lands were hives of horticultural activity 2,500 years ago, when sophisticated irrigation systems assured plentiful harvests in such a perfect climate for growing lettuce all year round.

With the fall of Rome, written accounts of cultivated lettuce hit something of a dead patch. Chaucer (1343–1400) wrote about a cultivated form of lettuce in rather unfavourable terms in *The Canterbury Tales*. Illustrations by Peter Schöffer (*c*.1425–*c*.1503) of four distinct types, none of which resembles anything we think of as lettuce today, were included in an encyclopedia of natural history compiled by Jacob Meydenbach in 1491 (botanical illustration was rather subjective back then). The first descriptions of cabbage lettuce – those with round, solid hearts – were made by Leonhart Fuchs (1501–1566) in his *New Herbal*, published in 1543. Other sixteenth-century botanists, including Rembert Dodoens (1517–1585), the French botanist Jacques Daléchamps (1513–1588), Pietro Andrea Matthioli (1501–1577) and the 'Father of German Botany' Jacobus Theodorus (1525–1590), describe various types of loose-leaf, cabbage and butterhead lettuce, with added descriptions from Greek and Roman commentators, including Pliny the Elder (23–79 CE) and Herodotus. The importance to our health and happiness, both spiritual and temporal, of 'salad' was passionately articulated by John Evelyn in his *Acetaria: A Discourse of Sallets* (1699): '...by reason of its soporigous [soporific] quality, lettuce ever was, and still continues the principal foundation of the universal tribe of Sallets, which is to cool and refresh, besides its other properties...including beneficial influences on morals, temperance, and chastity.'

A Recent Arrival

Various species of wild lettuce were first described in Britain by the botanist Thomas Johnson (1600–1644) in 1633.[5] *L. serriola*, the wild species that begat most cultivated lettuce, only firmly established itself across the land with the arrival of the Industrial Revolution in the eighteenth century. This is due to its weedy habit, meaning it thrives in disturbed soil. With the expansion of the canal network, which started in the middle of the eighteenth century, followed by the railways some 50 years later, the conditions for wild lettuce to flourish became much improved. The tendency of lettuce to grow best on disturbed ground is a great trait to have when it comes to evolution because, as a species, humans are very effective soil disturbers. Thanks to our distant ancestors becoming farmers, weedy plants have thrived around us. Lettuce lovers all have their favourites and often a love-hate relationship with the different forms. I will grow butterhead types for seed but please don't ask me to eat these soft, slippery things. My favourites are all traditional cos varieties.

Cultivated lettuce, introduced from mainland Europe, was firmly established in Britain from the reign of Charles I (1600–1649), although it wasn't until the middle of the eighteenth century that we find the British breeding their own lettuce. Brown Bath Cos, first recorded in 1743, is a firm favourite of mine, having been bred just a few miles from my home. There is a gene bank in the Netherlands, CGN, which holds a vast collection of lettuce. Their oldest named French cultivars are *Passion Blonde à Graine Blanche*, first recorded in 1755, and Palatine, recorded in 1777. Mescher, which originates from the Austro-Hungarian Empire, is another cultivar dating from the eighteenth century that was also being grown in England. For an overwintering butterhead it's actually very tasty. The changing landscape may have provided the ideal environment for wild lettuce to become a pretty addition to the flora of railway cuttings, verges and field margins, but it was in gardens that the lettuce we enjoy today was already feeling at home.

Spoiled for Choice

My staple go-to summer lettuce is the late-nineteenth-century, French-bred Little Gem. It's a cos type which also goes by other names. The Americans refer to it as having the crispness of a Romaine type and the sweetness of a butterhead. Romaine, or Roman Lettuce, is the name the French gave to this group of lettuces when several Popes moved from Rome to Avignon early in the fourteenth century, bringing it with them to grow in the palace gardens. For this reason, the Romaine is also sometimes called the Avignon Lettuce.

When it comes to flavour, Somerset breeders in the nineteenth century came up with a beauty, Bath Cos, which was first recorded in 1842 and widely grown as a commercial crop around the world. Another of my favourites is Lobjoits Green Cos. Dating back to the 1790s, this large variety can be grown all year round and is a superb overwintering salad crop. Bunyard's Matchless, first bred in Kent in 1855, is another fine, large, overwintering cos that I enjoy as much for its name as its flavour. High on my list of very yummy lettuce to grow through the winter is the loose-leaf or cutting type, Bloody Warrior, an ex-commercial British variety from the late nineteenth century, so called because the dark green leaves are splashed with deep red, almost black blotches. You won't find modern cultivars in my salad bowl, but those old-fashioned varieties that grow more slowly through the autumn and winter months and are at their best in early spring when their flavours are unimpeachable.

Changing Fashions

Popular today but also with a long culinary history are modern cutting lettuces. Those ubiquitous bags of tasteless leaves in supermarkets belie the fact that the Italians have been masters of producing wonderful ranges of cut-and-come-again lettuce, including a single variety, Lattughino, which in early spring delivers tangy, cut-and-come-again leaves followed by flavoursome crisp-head lettuces for Easter. American varieties with lovely names include a fine heirloom loose-leaf, Amish Deer Tongue, and the crisp, loose-headed, overwintering Bronze Arrow. It is testament to the quality and importance of these old varieties in our food culture that so many continue to be available commercially.

Lettuce was much written about in gardening circles throughout Victorian times. The 1844 edition of *The*

For the Love of a Leaf

Gardeners' Chronicle says of White Paris Cove Cos lettuce:
'This is unquestionably the finest Cos Lettuce that can be
grown for summer use. Though tall, the outer leaves are
hooded so as to close at the top and blanche the heart without being tied'. Oh, would that I could grow this culinary
wonder today but, sadly, like more than 90 per cent of all
our vegetable varieties, it has been lost for ever. Or maybe
not? The closely named White Paris Cos lettuce might be
the same. Still widely available in the U.S.A., I can but hope.
Writing in the same publication Mr R. Milne has this to say
on the subject of eating salads in England:

> 'Why do English gardeners so much neglect salads?
> On March 13th I came to Sidmouth, South
> Devon, an eminently favoured locality. The hotel
> is in a grand position, surrounded by choice coni-
> fers, such as Pinus insignis etc., with Camellias in
> flower, and a *walled-in kitchen garden of 7 to 8 acres
> with conservatories etc.* [my emphasis]. Here the
> only salad to be seen was Mustard and Cress, most
> ostentatiously served at every meal, as if a great
> delicacy; but no Lettuce, Endive or even Celery;
> no Radishes, nor fresh mint for sauce. How differ-
> ent from the north of France, where with a much
> worse climate, one has, at this time of year, Radis
> 'Rond à Bout Blanc' served each day at breakfast
> and delicious Lettuce salads at every meal. I came
> on to Falmouth, an even more favoured locality
> in the midst of private gardens, which not even
> the Riviera can equal – where one sees Tree Ferns,
> Palms, Embothrium, Dicksonias and other choice
> plants in full luxuriance. Here I am able to get

131

a little tough Cabbage-Lettuce and some even tougher Endive, most sparingly served, and that is all. Can it be that the cooks prefer to send to the greengrocer for French Lettuces? I am afraid there is something in this; but the hotel-keepers do not give these to their guests.'

TripAdvisor came 150 years too late for Mr Milne, I fear. No doubt such a discerning lettuce-eater would have wolfed down Webb's Wonderful, a fine, crunchy summer lettuce that has been a part of British food culture since those days, bred by a horticultural institution based in the West Midlands, Webbs of Wychbold. Another example of this is the crisp-head Chou de Naples, bred in France in the late nineteenth century and the parent to a large number of crisp-head cultivars sold by Vilmorin-Andrieux. However, it was the Americans who, inspired by the crisphead, took lettuce production to a whole new level.

The Tip of the Iceberg

Until the creation of the 'iceberg', lettuce was a local crop and didn't travel because of its poor keeping qualities. A heritage variety going by that name was the first to be bred in the US in 1894 and is still available today. It took until the 1930s and '40s for breeders in Salinas, California, to develop modern varieties from French Batavia types – a loose-leaf lettuce that has good keeping qualities – and the iceberg.

The US government was active in plant breeding from 1825, when President John Quincy directed US Consuls around the world to forward rare seeds and plants to the State Department for propagation and distribution. The

USDA was established in 1862 'to procure, propagate and distribute among the people new and valuable seeds and plants.'[6] Arguably the most important cultivar of iceberg is Great Lakes, the very first true modern iceberg lettuce. It was the result of a breeding programme by Thomas W. Whitaker of the USDA and the Michigan Agricultural Experiment Station. Despite the name, it was actually bred in California. Great Lakes started selling in 1941 and at one time made up 95 per cent of all iceberg production in the USA. It remained the market leader until the mid-1970s.[7] The first iceberg cultivars, when packed in ice, could be shipped to the East Coast by train and remain edible over days, even weeks.

The need for a long shelf life revolutionised lettuce cultivation, but in my humble opinion did nothing to improve the quality of our diet. Sadly, the modern iceberg, although ubiquitous in American food culture, is a tasteless addition to any sandwich, burger or salad. It is therefore with considerable relief that the Americans are finally seeing the error of their ways and turning to better fayre. Now, slightly more flavoursome, but still pretty disagreeable in my view, pre-packed, loose-leaf lettuce takes number one spot. It is also refreshing, both figuratively and literally, to see many traditional varieties as well as modern hybrids filling the salad aisles of the better-quality supermarket across the land. The ghastly iceberg, like its chilly namesake, is melting away, although a large, cold wedge drenched in sweet, thick dressing remains a staple of many Americans' diets.

For those of us who like to save seeds, lettuce is a fairly easy crop because it is self-pollinating.[8] This is not to say that lettuce will not cross-pollinate. If one is growing just a few lettuces, then about 5 per cent of the crop will hybridise with its neighbours up to about 40m (130ft) away. This ability to cross, especially with close relatives – as happened in the early days of domestication – is the reason why the lettuce has been so successfully domesticated over the last 5,000 years or so and is also key to its long-term survivability. Even with just 5 per cent gene-sharing, the lettuce remains genetically diverse and robust.

Although the flowers are self-fertile and do not secrete much, if any, nectar they are very attractive to various pollinators, including honeybees and hoverflies. Yet, one might ask, if lettuces have evolved so that they don't need pollinators to procreate, why have they remained attractive to them? The answer is quite simple: in the process of evolutionary change and adaption many genetic traits become redundant. However, these genes, although not needed, remain within the plant's genome as they are not inhibiting its ability to survive.

A Blue-Flowered Cousin

Lettuce, for all its shapes and sizes, its seasons, its place in our culinary universe, and its importance in our social and cultural identity, is but one of the many salad crops we now consume. Today we eat leaves and shoots from plants that were never originally regarded as something you could make a salad with: microgreens of beetroot, spinach, brassica leaves of all types, pea shoots – the list seems endless. But there are two salad crops that are eaten as leaves, both raw and cooked, one of which also has a large root that makes a particularly filthy alternative to coffee. Of the genus *Cichorium*, they are chicory and endive, my absolute go-to autumn and winter vegetables. Frequently treated as a single type of vegetable – confusingly the French call chicory 'endive' – they are closely related but distinct species. For truly tasty leaves outside the summer months, it is really thanks to breeders from Italy, France and across Northern Europe that we have such an exceptional number of varieties to choose from.

Sixteenth-century herbalists used the Syrian word *ambubaia*, which refers to the thin and hollow shape of the plant, to describe wild chicory, *Cichorium intybus*. It is native to Europe, the Middle East and much of North Africa, and its bitter leaves were edible – just. We have the Syrians and Egyptians to thank for developing a more edible variety of domesticated chicory at least 4,500 years ago. In England, wild chicory was traditionally known as blue succory on account of its lovely, blue, dandelion-like flowers. Its common name in France was *chicorée sauvage*. The French also called a cultivated version of this charming plant *Barbe de Capucin* (Capuchin's Beard).

From Short-Lived Secret to Longer-Lived Drink

Traditionally, chicory's bitter leaves were blanched to make them more palatable, a method that continues to this day, as evidenced by those pale green, white and pink 'chicons' we now find aplenty in supermarket chiller cabinets. So how did we arrive at those perfectly shaped specimens? The story goes that in 1844 a Belgian gardener, who was experimenting with growing wild chicory for its roots, discovered some white leaves coming from a plant buried in a mushroom bed – an accidental mutation of wild chicory. Thus was born the famous Witloof or Large-Rooted Brussels chicory we consume today. *Witloof* is Flemish for 'white leaf', which is also an old name given to wild chicory. Our nameless Belgian tried hard to keep his discovery secret, but such was the nosiness of his fellow gardeners, or more likely the extent of his bragging, that soon everyone was growing the plant and it became a much- prized and highly valued commercial crop.[9]

Chicory root was first used to make a coffee substitute in 1779, three years after Prussia's Frederick the Great banned the import of coffee beans. An innkeeper in Brunswick, Lower Saxony, was granted a concession by the monarch to manufacture a substitute which became the basis for Camp coffee, a sickly syrup invented in Scotland in 1876. The use of chicory as a coffee substitute leaves me cold. It may have been all very well back in the day when coffee was expensive and in short supply, yet it became a stalwart of British wartime beverages. Growing up in the 1950s and early '60s, I viewed a jar of this black goo with contempt. It took just one sip of my mother's brew to know that this was not a drink for me. It continues to be manufactured to this day and it amazes me that anyone would still want to drink it! I believe that a chef in North Yorkshire is growing chicory to make a new coffee alternative, so perhaps I should put my childhood prejudice to one side and see if he can change my mind. Maybe a tastier alternative is a traditional tea of Korea that is made using the dried chicory root.

The other species of *Cichorium* that is so significant in our food culture is endive (*C. endivia*). It is the product of domestication of its wild parents *C. pumilum* and *C. calvum*, both native to the shores of North Africa. It might have been cultivated by the Egyptians, although there is no archaeological evidence to support this idea. The earliest mention of endive is by the Roman poet Horace (65–8 BCE) in one of his famous odes listing the ingredients of his simple diet. He would have been using the word endive to describe a domesticated form of wild chicory because at the time no clear distinction between the crops was made. Like chicory, it first appeared as a cultivated crop in northern Europe during the sixteenth century and now comes in many delicious guises.

Today the most popular types of endive are divided into three cultivar groups. First is the curled-leaved group *C. crispum*, giving us open-hearted, frizzy endives, which have their centres blanched by excluding light. In the middle of the nineteenth century, the great French plant breeder Vilmorin-Andrieux was selling literally dozens of varieties of this group known as *Frisée*, including my favourite, Chicorée Frisée Grosse Pancalière.

Escarole is the French name for the second group *C. latifolium*, broad and flat-leaved, open-hearted endives which are also sometimes blanched. Hard to beat for flavour are French classics Géante Maraîchère and Grosse Bouclée. The third group, *C. folosum*, are Italian varieties of radicchio. Classics that have been cultivated since the nineteenth century include Grumolo Rossa and Rossa di Verona – which can be blanched like Witloof – and red and white varieties like Variegata di Chioggia. There are Catalan chicories of the same family group as radicchio too, which do not form solid hearts, like Puntarelle di Galatina. Needless to say, this wonderful pantheon of green, red, variegated and white crops has taken centre stage in much European cuisine, and they are now grown around the world.

In Britain, botanists failed to distinguish cultivated endive and its kind from blue succory until the twelfth century, and for the next 400 years it was not generally eaten as a salad crop. Whitloof chicory was the first cultivated form of the genus to be eaten in the U.K. and it was not until the latter part of the twentieth century that we embraced this awesome vegetable in all its guises. Long may this continue.

There is something very reassuring about cruising the markets and quality greengrocers to see an ever-increasing number of different salad crops for sale. For the most part they are modern cultivars, bred like everything else to grow uniformly and quickly, to have a long shelf life, and to present well to the shopper. So often flavour has been sacrificed in the breeding programme but there are many stalwarts that keep tickling one's taste buds. Lettuces like Little Gem, more and more open-hearted lettuce, and endives with colour and texture are becoming increasingly popular too. Supermarket aisles are filled with radicchio, red and white chicory, the heavenly escarole – all nutritious and tasty. However, salad leaves packed in nitrogen-filled bags to keep them 'fresh' for longer are not only stupidly costly, but also disappointing for flavour. We should not abnegate the sink, a sharp knife and a salad spinner. Forsaking such for convenience and laziness is to miss out on a truly worldly pleasure. Come rain or shine, in summer heat or winter chill, having lettuce, chicory and endive in all their myriad forms always growing in my garden, to be harvested every day of the year if so desired, is simply magic.

Thank Goodness for Garlic

If any man with impious hand has broken his father's neck,
Let him eat garlic. It is worse than Hemlock.

Horace (65–8 BCE) – *Epode III:*
That Wicked Garlic (*c*.30 BCE)

I t was another glorious summer's day. The end of June and, as I opened the front door, returning home after a week's holiday, I was met with a wall of perfumed air. Garlic. Pungent and earthy. I followed my nose into the kitchen and awaiting me on the table was a large plastic sweet jar completely covered in a riot of colourful stamps. It took but a moment to remove the sellotaped lid to reveal one of the most heavenly sights a garlic lover like me could wish for: the jar was packed to the very top with large, fat, deep purple individual cloves of a vegetable I eat almost every day. I was amazed…and then I remembered.

A Land Full of Surprises

Back in March I had been in Oman. The first place I visited on my arrival in the capital, Muscat, was the Central Vegetable Market on the edge of town. A huge, modern complex of warehouses, it was home to a vast range of fruit and vegetables, many of which were locally grown. Sacks of carrots labelled Desert Magic, an endless variety of cucumbers, beans, herbs, bananas, mangoes, tomatoes and, of course, fresh young garlic. Oman is famous for its dates and there was no shortage of them being sold as well. Having arrived here knowing nothing about its long connection with cultivation I did not think on that balmy morning, strolling through the market chatting to the traders, that this desert kingdom had a vibrant culture of local varieties. Indeed, many farmers, with the encouragement of their government, are growing and saving seeds of a number of crops, the most important of which is the fodder crop alfalfa. They have been saving their own seed for generations and, as a result, Omani alfalfa has become superbly adapted to growing in a climate where summer temperatures can top 50°C (122°F).[1]

It was time to explore a little further. Next stop, Jebel Akhdar, a mountainous part of the Saiq Plateau. At about 2,000m (6,500ft) above sea level, this stunning region is famous for growing damask roses to make rosewater, possibly Oman's least-known but loveliest export after its sensational dates! When I travelled there in 2012 the road had only recently been partially tarmacked. A four-wheel drive was essential and a smile at the checkpoint off the main highway was sufficient to allow free passage. Tiny, ancient villages cling to precipitous cliffs as terraces tumble 1,000m (3,280ft) to the valley floors. As well as roses, all just starting into bloom when I was there in early spring, farmers grow pomegranates, peaches, grapes and a profusion of vegetables, including local varieties of tomato, cucumber and garlic. As I wandered along the maze of narrow paths and tracks that connect the villages and their terraced gardens, I saw young garlic plants that had been growing through a winter where there are frequent frosts but little rain. First populated by the earliest Arab settlers in Oman known as the Bani Riyam, it is no wonder that over hundreds of years the indigenous farmers built some of the most magnificent terracing one can find anywhere in the world. With just over 30cm (12in) of rain a year, around half that of the driest parts of eastern England, every drop of water is precious. It is collected through a sophisticated network of irrigation channels and holding tanks built over centuries, meaning there is enough water to grow pretty much anything. With short, chill winters, oodles of sunshine and temperatures that rarely climb above 31°C (88°F) even in the hottest summers, roses are not the only plants that are very happy growing on the plateau.

Oman has always been part of the ancient coastal trading routes between north and east Africa and the Indian subcontinent. Archaeological evidence at El Wattih, outside Muscat, shows that Oman has been inhabited from the times of the Fertile Crescent's earliest settled communities some 10,000 years ago, so its agricultural heritage goes as far back as anywhere on Earth. Seeds were carried by people travelling back and forth, both by sea and along the ancient trade routes to Aleppo and Constantinople. This ensured that for thousands of years there would have been a wide range of fruit and vegetables being grown in that fertile corner of the Arabian Peninsula.

Returning to the hotel at Al Jabal after discovering garlic growing nearby, I asked our host Nabhan Al-Nabhani if they were a local variety. With a broad grin he told me that he owned the garlic terraces I had been walking. His family had been growing the same garlic on them for generations. He thought it might have been introduced to Oman by the Portuguese, who controlled the region throughout the sixteenth and first half of the seventeenth century. This may well be true, but my seed-detecting nose started twitching. It would also have arrived in Oman with traders as well as Babylonian and Assyrian invaders intent on controlling the trade routes over 3,000 years ago. If so, this garlic had probably been growing in Oman for millennia. I dared to ask Nabhan if he might send me a few cloves when they were ready to harvest in the summer. He promised he would, which is why there was a fragrant jar waiting for me on the kitchen table.

Importing plant material is, or should be, strictly controlled. Alliums, which include garlic, are no exception. Looking at the stamp-covered sweet-jar on my kitchen table, it was clear it had been assessed by customs and allowed to continue to my door; I did not care to make further checks. However, with a kilo of individual garlic cloves demanding my attention, the first job was to see just how tasty they might be. I was not to be disappointed. When removed, the garlic's deep purple outer skin revealed a lovely white clove with the slightest purple blush. The flavour was exceptional. Pungent and bold is how I like my garlic, and this treasure from Oman did not disappoint. I gave a few cloves away to friends and neighbours, selected about forty to grow the following year and ate the rest.

The first task facing me when presented with or discovering seeds, or in this case bulbs, from very different climatic regions is to see how well they take to growing in my garden. In fact, nearly all the fruits and vegetables I cultivate are remarkably good at adapting to our weather here in the U.K. Having a polytunnel helps. I knew that the Omani garlic had been planted in the autumn and was ready for harvesting by June. Most garlic need a period of cold to trigger bulb development, otherwise all you get is one, large single clove. Also, the Saiq Plateau gets rather chilly, with regular frosts between December and February, but it is a dry time too – a far cry from our chill, damp, long winters. So, this garlic would have the best chance of success being grown under cover. It proved to be an exceptional 'keeper', considering it was ready for harvesting in early summer. A 'hardneck' type (more of this anon), it was at its best eaten before the end of the winter.

A Healthy History

Of all the vegetables in this book, garlic probably lays claim to more health benefits than any other. Like its distant cousins, the leeks and onions, it is a member of the *Allium* genus in the Amaryllidaceae family. Research undertaken early this century points to the centre of origin for garlic being on the northwestern side of the Tien Shan Mountains of Kyrgyzstan and Kazakhstan.[2] Among the earliest cultivated crops, it would also have been foraged long before Neolithic farmers selected the best examples to plant at least 6,000 years ago. The oldest garlic remains have been found at the Cave of Treasure close to the Dead Sea, near Ein Gedi in Israel. They have been dated to the early fourth millennium BCE. It was widely grown throughout Central and Eastern Europe and North Africa 5,000 years ago.

Garlic was much valued by the Egyptians, particularly the workers, and given to slaves building the pyramids as a daily ration because of its alleged ability to fortify them. Garlic was found in the tomb of Tutankhamun (*c*.1342–1323 BCE). Whether left there by design or accident is not known. There is a rather engaging story told about the Greek historian Herodotus (*c*.484–*c*.425 BCE) who, on visiting the Great Pyramid at Giza, wrote of the exorbitant price the pharaohs had to pay for garlic and some other vegetables to feed the workers. He probably misunderstood, or something was lost in translation, because what was very expensive at the time was a type of arsenate stone used to build pyramids. When burned it smelled of garlic.[3]

The Talmud suggested that garlic be used as a way of dealing with stomach worms and parasites and as an aid to procreation – the vegetable as aphrodisiac yet again! There is no evidence in Jewish or Egyptian texts of garlic having a religious significance, however. The Greeks attributed to garlic the same qualities as the Egyptians and fed it to soldiers to make them fight better. It was also prescribed to the first Olympian athletes, making it the world's first performance-enhancing drug.[4] One wonders if there were limits to the amount consumed by ambitious athletes – might there have been prescribed doses, clinical trials even?[5] Nero's chief physician to the army, Dioscorides (*c.*40–*c.*90 CE), thought garlic was good for the arteries – although he believed these carried air, not blood. He may have been on to something, however, because garlic's potential as a cardiovascular drug is the subject of considerable research today. Pliny the Elder (23–79 CE) listed 23 disorders that could be treated with garlic. The ancients certainly did know a thing or two about this vegetable's undoubted medicinal properties.

The first written accounts of garlic as a medicinal herb appear in Sanskrit documents from 3,000 BCE. At the same time as the Greeks and Romans were feeding it to their workers and soldiers, Indians were using it to treat heart disease and arthritis. Introduced to Western Europe by the Romans, it was used as a remedy for tuberculosis, plague and cholera, as well as all the other ailments already mentioned.

The Chinese call garlic suan. This is written as a single character, which suggests the word to be very old – pointing to garlic having a significant role in ancient Chinese food culture too. Four thousand years ago, the Chinese used garlic as a food preservative and liked eating it with raw meat. Again, they recognised garlic's use for treating intestinal worms, and it was also used in combination with other herbs to treat depression, impotency, insomnia, headache and fatigue.

The vegetable's Latin name *Allium sativum* is apparent in the French *ail* and the Italian *aglio*. The word garlic comes from *garleac*. *Gar* is Middle English for 'spear' and *leac* is Anglo-Saxon for 'herb', describing the plant's spear-headed clove.

A Food for Good and Ill

Until Henry IV of France had garlic added to his baptismal water in 1553, allegedly to protect him from evil spirits, it was seen as peasant food, only to be eaten by the poor and working classes. However, as a medicine it was consumed by everyone. Doctors would carry garlic to protect themselves against the odour of disease – and, of course, any self-respecting vampire hunter needed to keep their garlic supply close by. Although we tend to associate vampirism with Slavic countries and cultures, vampires were present in pre-Christian European cultures, within Judaism, and across Mesopotamia. Garlic was used to ward off vampires in China and many parts of South-East Asia long before it became the go-to protection in European culture.

However, not everyone is enamoured of garlic. Some Buddhists exclude the five pungent alliums: garlic, onions, leeks, chives and spring onions from their diet because they believe these have an adverse effect on our bodies. Brahmins and Jains do not eat alliums either. According to the ancient Indian medical text, the *Ayurveda*, foods are grouped into three categories: sattvic, rajasic and tamasic, representing goodness, passion and ignorance, respectively. Onions and garlic are classified as rajasic and tamasic, which means they increase passion and ignorance and should be excluded from one's diet.[6]

The chemical compound that gives garlic its distinctive aroma is allicin. This sulphur-containing ingredient, which is found in all edible alliums, is at the heart of the many medicinal qualities of garlic. The internet is packed with sites extolling the health benefits and medicinal virtues of garlic, with any number of lists of things the bulb can cure. Clinically proven attributes are as an anti-inflammatory, an antiseptic, and an antifungal. During World War I it was used on wounds as an antiseptic and to reduce dysentery in the trenches. Much research has been and continues to be undertaken to see if its consumption can reduce cholesterol, blood pressure and the likelihood of contracting certain types of cancer, including lung and brain cancer. It's also quite effective at removing warts apparently. However, reading the small print of many health food and alternative medicine products, medicinal attributes are more often 'believed' or 'claimed' than clinically proven. Garlic 'may' have antibiotic properties, it 'may' help in weight reduction, it 'may' help fungal skin infections and, like so many other foods, it is claimed to be an aphrodisiac because it improves circulation. What is certain is that garlic is the focus of a wide range of ongoing clinical research and trials, and eating it regularly and in quantity may not make you popular, but it will almost certainly be good for your health.

Garlic is nature's greatest superfood and one of the most important weapons in an herbalist's arsenal. It is an ingredient I really cannot imagine being without and it has been arousing passions of a culinary nature in food cultures around the world for millennia. I witnessed a prime example of this recently while travelling in Rajasthan in northwest India. On the trail of threatened local varieties of edible crops, I was foraging in a market in the fortress city of Bikaner where piles of locally grown garlic were being sold. Indian garlic is unlike the more blousy modern Chinese varieties, which are large and have fewer big cloves than their pungent, pale Indian cousins. The people of Rajasthan are justifiably proud of their garlic, where each bulb which can contain twenty or more thin cloves. No self-respecting Rajasthani would be seen dead buying Chinese garlic, as was made evident to me by both an impassioned stallholder and my farmer guide. Garlic has been a staple of Indian cuisine for at least the last 4,000 years. Although now very different to the purple beauty I discovered in Oman, it is closely related for reasons I will expand on later.

A Seedless Story

Garlic comes in two forms: softneck *Allium sativum* (so called because as the plants ripen the stems flop over), which produces smaller bulbs with many tightly packed cloves such as the Rajasthani garlic, and hardneck, *A. sativum* var. *ophioscorodon* (so called because it remains upright as it ripens) – the garlic from Oman being an example. Although garlic was growing wild across Central Asia, Indo-China, North Africa and Central Europe at the time of domestication, the wild parents of most garlic can now only be found in the colder regions of garlic's centre of origin in Central Asia. Unlike softnecks which are sterile, many hardnecks produce flower spikes known as scapes. These are delicious when eaten before the flower head opens. However, the flowers that form are almost always sterile, containing only bulbils and no seeds. These hardneck varieties produce fewer and larger cloves with a less pungent flavour, which is what the Rajasthanis turn their noses up at! They don't keep as long as softneck varieties either and are planted in the autumn because they need cold weather to trigger clove production. They are absolutely delicious when harvested as immature bulbs in early summer and can be bought as 'fresh' garlic. Softnecks are completely sterile, rarely producing flower stalks and, from an autumn planting, mature in mid-summer, when they can be stored through to the following spring.

Since garlic was first domesticated it has been propagated by one method: division. You take a single clove and, when you plant it, you end up with a bulb of many cloves, all of them genetically identical clones. Some hardneck types produce flowers and seeds but propagating from them, let alone developing new cultivars, is complicated and not for the faint-hearted, as we shall see.

Wild garlic, *Allium ursinum*, also known as ramsons, which we see growing in open woodland in spring with its drifts of lovely white flowers and glossy green foliage, is a completely different species, although of the same genus, and should not be confused with cultivated garlic. It is the familiar and pungent smell that ramsons and garlic share – hence the common name. It is also much sought after by foragers supplying trendy restaurants at handsome prices, but for those of us prepared to wander in open woodland in spring, this delightful and delicious herb can be had for free.

The world of academia can be an argumentative one. There is continuing discussion as to who garlic's true wild parent might have been. One group of researchers believes it could be *A. longicuspis* Regal – named after Edward August von Regal (1815–1892) who described the plant, which is indigenous to parts of Central Asia, in the middle of the nineteenth century. It has been found by plant hunters in the rather wonderfully named Heavenly (in Chinese, Tien Shan) Mountains on the China-Kyrgyzstan border and would have been grown there more than 6,000 years ago. By selecting only certain bulbs, seed production was inhibited in favour of more and bigger bulblets or cloves, as today this variety is presumed to be sterile. But is it?

Despite garlic's most illustrious and well-documented history, it is only very recently that it has been described in varietal terms. Plant breeders and geneticists have debated for some time as to what type of garlic the first farmers living beyond its centre of origin would have grown from selections they had taken from the wild varieties. In writings dating back 5,000 years, garlic is never referred to by colour, size or habit. It was only when it began to be cultivated in southern Europe a thousand years ago that distinctions between hardneck and softneck were noted. Philipp Simon, who is something of a garlic guru at the Department of Horticulture at the University of Wisconsin, was lucky enough to collect wild garlic in Central Asia towards the end of the last century. He found mostly hardneck specimens and some softneck. The hardneck plants were prolific producers of seed and this led him to conclude that the considerable diversity of colour, shape and flavour we find in cultivated garlic grown over the last millennia is thanks to considerable genetic variation brought about as a result of cross-pollination and hybridisation among the wild populations. Ongoing research proves that some genotypes of cultivated garlic do, in fact, produce seed. This is now the subject of further work to breed new varieties that can be improved through conventional sexual breeding techniques.[7]

As far as Simon and other academics know, garlic was only ever reproduced asexually in cultivation, just as we do today, by separating out the cloves from a single bulb and planting them. So, garlic has not changed much over several thousand years, including the garlic I found in Oman and Rajasthan: they are all clones and genetically very uniform. Well…almost.

Plant hunters and botanists have identified a number of other varieties of garlic that grow in specific parts of its native homelands of Central and East Asia. An example is *Allium tuncelianum*, which is endemic to the Munzur Valley in the Anatolian region of central Turkey. It is a prized part of the local food culture, but sadly, this garlic is highly endangered due to over exploitation and problems of poor seed-set. It is the subject of much research to secure its continuing survival through in-vitro and other propagation techniques. For many years it has been grown in European gardens as an herbaceous flowering bulb rather than from seed. Recently, it has attracted the attention of breeders because it is also a fine culinary garlic with very high levels of allicin. It flowers freely, sending up lovely curly scapes that produce large, ball-shaped, pale purple heads which ripen to yield quantities of little black seeds. Growers are interested in it because garlic can be infected by viruses. Because they are propagated asexually, these viruses lodge in affected plants, reducing the quality and size of the harvest. But with seed, this is not an issue. Any virus present in the clove stays in the clove, so the seed is disease-free. Great too for plant breeders, who can develop new varieties using sexual reproduction.

A Very Modern Plant

The big challenge for breeders has been how to improve varieties through sexual reproduction by cross-pollinating from a number of different clones. A frustrating trait of domesticated garlic is that any flowers that do form rarely produce seeds. A typical flower contains tiny bulbils which are simply more clones of the parent, and these crowd out the flowers and prevent them from developing. Breeders have to remove these bulbils, so the few flowers that remain can be pollinated. The plants must be grown in carefully controlled climatic conditions to reduce bulbil proliferation. This requires patience because the breeder needs to select from the sexual-ly-reproduced offspring specimens that stop making bulbils and focus on delivering true garlic seed. This way of producing new cultivars only started in the 1980s. Before then just a few thousand clones would have been grown over millennia.[8]

In 1991, the agricultural geneticist Alessandro Bozzini wrote a paper in the scientific journal *Economic Botany* about a wonderful discovery he had made in 1987. Growing in a field near the small town of Caiazzo some 30 miles north of Naples was a local garlic that produced flowers contain-ing seeds. He took some of this seed back to his lab at the University of Naples and was able to germinate 80 per cent of them – a remarkable rate of fertility.[9] Now, perhaps for the first time in history, plant breeders are able to work with crossing and hybridising garlic sexually to create new varieties that can enrich our diet. DNA fingerprinting is also going to help us better understand garlic's origins and its distribution.

From Hate to Love

The British had a unique distrust and dislike for garlic. Even today, some older generations abhor it. There were xeno-phobic undertones regarding garlic lovers, from southern Europe particularly, who suffered from 'garlic breath'. As a teenager I was expected to brush my teeth after a meal that included garlic to eliminate any foul personal odours. It's a different story today. The vast majority in my homeland embrace garlic, thanks, I am sure, to the positive influence of 'foreign food' in the last 50 years. Now seed catalogues offer dozens of varieties, most grown by specialist U.K. farmers. Just a few years ago we imported all our garlic, both to grow and to eat. No more. I share an obsession among gardeners, growing new and ever more exotic varieties.

Recently I was giving a talk about one of my seed-detecting adventures to a gardening group in Nottingham, a city not short of allotments – there are over a thousand. The venue was the Polish Club, a large room with a bar at one end. As I gave my talk, I could see brothers Marian and Jack Knapczyk leaning on the counter listening. When I had finished, I was invited to the bar to savour some good Polish beer and handed a fistful of fat cloves of a pink Russian garlic. They had been grown by a nameless Russian émigré who had brought the garlic with him from his homeland some years ago. He would share these out in the Polish Club every year. Marian and Jack were not gardeners, so didn't grow any themselves – they didn't need to as they were regularly supplied! They were great garlic fans, however, and we had a lengthy and somewhat esoteric conversation about what constituted good garlic. Needless to say, I planted those cloves and have now added them to my garden plan. They form large hardneck bulbs, which in May send up flavoursome scapes as curly as pigs' tails. In all my travels through the former Soviet Union I never came across such a fine pink garlic, so it is a thrill for me to have them, plaited and hanging in the porch, a culinary delight into winter. My small collection of garlic varieties is testament to the richness and diversity of our relationship with this magic allium. Something to be treasured.

PART TWO

Arrivals from the West

Excavations undertaken by Richard MacNeisch in the 1960s in Mexico's Tehuacán Valley provides us with a brilliant timeline that shows how our South American ancestors moved from being hunter-gatherers to farmers. Across some 12 sites MacNeisch and his team found evidence of agriculture dating back some 12,000 years. At the outset hunter-gatherers lived on a diet of foraged plant food and prey such as rabbits, small deer and lizards. Edible plants were systematically foraged as they came into season. It seems that about 9,000 years ago there was less game to be had – possibly because of over-hunting – but there could have been other environmental factors at play too. As a result, more time was spent collecting wild foods such as avocados and squash. The people worked in small foraging groups during the dry season when food was scarcer, joining forces during times of plenty. Any sporadic growing of wild plants would have been minimal.

Domestication of plants increased gradually over a period of 5,000 years, so that around 7,000 years ago, 10 per cent of our diet came from cultivated crops. Many of them had been introduced – their centres of origin were outside the Tehuacán Valley – including amaranth, maize, squash and chillies. For the next 4,000 years the people of this valley became ever-better farmers, until about 3,000 years ago when people started to grow almost everything they ate.[1] And then, just over 500 years ago, the wonderful diversity of domesticated crops that had evolved as a result of the genius of the indigenous people of South America found their way into the food culture of the 'Old World'.

The geographical descriptions of New World and Old World are a European invention, but the terms have become linguistically ubiquitous as shorthand. In truth, Europe at the end of the fifteenth century believed a 'New World' had been discovered. The fact that it was anything but new to the indigenous peoples of all the Americas should never be forgotten. Nor, when reminding ourselves of the glorious culinary discoveries that subsequently enriched all our diets, should any of us forget that they are a product of the violent displacement and genocide perpetrated on the people of the New World by European colonisers and the abduction and enslavement of Africans which started early in the sixteenth century in what became known as the slave trade.

Christopher Columbus (1451–1506) was charged by Queen Isabella of Spain to do three things: first, to find a direct sea route to India; secondly, to bring wealth back to Spain; and, finally, to convert the indigenous people he encountered to Catholicism. He reached an island in the present-day Bahamas in 1492, where he kept a detailed journal of his expedition, though by his own admission, a botanist he was not. The indigenous people, the Arawaks, introduced him first to tobacco, which they smoked through their nostrils using a simple hollowed-out piece of wood.[2] The Arawak subsisted on two sources of carbohydrate: a variety of cassava, *Manihot esculenta*, which grew in many parts of the world and was already known to Europeans, and the sweet potato, *Ipomoea batatas*, native to parts of South America.

Returning to Spain in 1493, Columbus carried with him maize, tobacco, sweet potatoes, chillies and two species of bean. Twenty years later, with the Spanish conquest of Mexico and Central America by the conquistador Hernán Cortés (1485–1547), more crops were introduced to Europe, including long fibre cotton, the avocado, pineapple, cocoa, squash, other bean species, tomatoes, cassava and the potato.

The Vegetable Influencers

Today we may think of the contemporary cultural imperialism of American fast food being manifest in the proliferation of hamburger joints. Yet the most profound change in the global diet began over 500 years ago, after Christopher Columbus stumbled across the New World in 1492. I find it extraordinary that so many vegetables that are barely given a second thought by Europeans, so embedded are they in our own food culture, were first farmed in a small corner of southern Mexico. Tomatoes, chillies, maize, and numerous different types of beans and squash are all native to that part of our planet and are fundamental to my own sense of self. Indians too, for example, have the Portuguese to thank for providing an ingredient that would revolutionise their cuisine: the chilli. Yet, on my travels in that country, no one I met knew that their favourite spice came from Mexico.

Columbus's journey across the Atlantic to the New World was a seminal moment in the transformation and globalisation of Western food culture because it introduced the Old World to a large number of crops that had never been seen there before. Their very existence challenged the prevailing orthodoxy of botany and a belief that all knowledge could be found within the writings of Greek philosophers. The vegetables in the following chapters were brought to Europe from the Spanish and Portuguese colonies in the New World in the first quarter of the sixteenth century. They caused a revolution not only in the European diet, but also in scientific thinking. The world was changed forever.

More Than Just a Fruit

*'This rubicund youth with his blunt
features, appeared for all the world
to have a tomato instead of a head'*

Marcel Proust (1871–1922) –
Sodom and Gomorrah (1921)

I t was a remarkably inclement summer, I remember, but the vineyards and small farms around Montepulciano in Tuscany were burgeoning with ripening vines. I was producing a TV series about a just-married couple, Englishwoman Katie and her Tuscan husband Giancarlo, who were running a cookery school outside the pretty town. Giancarlo, a talented chef with a successful restaurant in London, was keen to involve his Italian family in the making of the programmes. It was his cousin Nello whom I instantly recognised as a kindred spirit. In his bountiful garden was a fine plum tomato, tipped with a small nipple, that had been grown by him and his family for generations – a genuine Tuscan heirloom. Naturally, I was keen to take some seeds home to grow myself, but I was also curious to know how his tomato stacked up against similar types from other parts of Italy. Well, the tirade of insults aimed at the inferior nature of all other Italian tomatoes took me somewhat by surprise. I barely got out of his house alive. To have the temerity to suggest that a Sicilian tomato might be superior to a Tuscan one was to invite violent retribution! Needless to say, Nello's Plum lives up to its reputation and I grow it most years. Its provenance goes back to the start of tomato growing in Europe, which began 500 years ago.

Of all the crops in my garden, it is tomatoes that attract the most powerful feeling of ownership and cultural identity wherever in the world I find them. They form a large part of my library, some 70 varieties from across Europe, Africa, the Middle East, South-East Asia and the USA. Whether conserving ex-commercial British varieties, growing out seed to share or trying other gardeners' favourites just to see how good – or awful – some of them can be, my discussions with growers around the world always turn to the uniqueness and cultural importance of local varieties.

Colour Can Be Deceiving

Although modern cultivars mean that there is more choice, today's tomato is bred primarily for staying power. What you buy has traits that ensure a long shelf life, a uniform appearance, an even colour and claims for taste on the label, which are often a great work of fiction. There are exceptions. I grow one tomato called Sungold, which really is delicious. However, there is now a powerful movement among food writers and chefs to promote traditional heritage and heirloom varieties as well as a considerable hunger among consumers to buy them. Tomatoes coloured green, yellow, golden or white can also be found in supermarket aisles alongside the regular red ones, with claims of being heritage or heirloom. But, beware. Most of those little cartons of coloured and cutely shaped fruit are almost without exception modern hybrids.

Thankfully, armies of passionate gardeners, organic farmers and traditional breeders continue to maintain genuine heritage and heirloom varieties.

Even in buttoned-up Britain, growers like me argue endlessly about which varieties have the best flavour, sweetness, acidity and perfume. There is no doubt that the loveliest tomatoes are grown in countries that are warm and sunny, which is why it can be a challenge to grow really great tomatoes in the U.K. where a poor summer of low light levels and cool nights can lead to disappointing harvests. I should know; I've had plenty of bad years. High-tech solutions such as growing under artificial light using different coloured LEDs to promote better flavour and vigour are being embraced by commercial growers. But for me the only tomatoes worth eating are those I grow warmed by the sun and harvested when fully ripe. If I don't have any tomatoes growing in the garden, then I would prefer to go without than waste money on a pale imitation.

A Poisonous Family

The story of the tomato, *Solanum lycopersicum*, like so many of the vegetables and fruits we eat, is a colourful one. It is a member of the 3,000-strong Solanaceae or nightshade family, which includes the tree tomato or tamarillo, the humble spud and the aubergine (or eggplant as it is known in the US). Peppers and chillies are also members of this family, as is tobacco, the delicious Cape gooseberry – with its papery lanterns that are home to sweet, golden, cherry-sized fruits – and the tomatillo (one of my least favourite vegetables). Less savoury relatives include the poisonous deadly nightshade, the magnificent but just as poisonous flowering shrub

Datura and petunias – which I wouldn't fancy eating either. The hallucinogenic mandrake is another relative and, like the tomato, is full of alkaloids, some of which are deadly. When one takes even just a cursory look at the breadth of plants within the huge Solanaceae family, it is clear that it includes not only some of the most delicious of all edible crops, but also others that will kill you for sure. Unripe tomatoes contain the toxic alkaloid tomatine, which has fungicidal, insecticidal and antimicrobial properties that the plant needs to protect itself from predators and disease. Tomatine is broken down when cooked, which is presumably why dying from eating an excess of fried green tomatoes cannot be blamed on the plant. Because of its unripe toxicity, as well as its association with poisonous plants of the same family, the tomato had something of an inauspicious start when it first arrived in Europe. As we shall see, it was more than a century after its introduction before the locals viewed it with anything other than deep suspicion.

All the cultivated tomatoes that we enjoy eating today are descendants of the red-fruited wild tomato, *Solanum pimpinellifolium*[1] which is indigenous to the coastal regions of Peru. All wild tomatoes have some remarkable traits because the climate of their homelands can be challenging. With little rainfall except when El Niño is active, many wild tomatoes grow where the cool sea causes coastal fogs laden with moisture. Along with heavy dews, this climate provides sufficient water for them to flourish; others are found inland and often at considerable elevation – above 2,000m (6,560ft). A notable example is *S. chilense*, a drought-adapted species that obtains water through its deep roots. Wild tomatoes have greenish fruits, with the exception of the red-fruited *S. pimpinellifolium*, and in general they are weedy plants.

Of the 14 species of wild tomatoes indigenous to South America, one has found a particularly remarkable niche: *S. cheesmaniae* is endemic to the Galapagos Islands and is a much-favoured snack for the other local inhabitant, the giant Galapagos tortoise.

Tomatoes are an opportunistic fruit too because they can grow both as a perennial with sufficient moisture, but also as an occasional annual in times of drought. The seeds can survive long periods of desiccation in the ground. Tomatoes are also very good at recovering from wilting if deprived of water at the height of their growing season, and their seeds are pretty much indestructible when composted or digested. Even passing through a sewage works won't kill them. I well remember as a kid on the farm being delighted to see the mounds of processed human effluent delivered to us from the local sewage works as fertiliser turning, chameleon-like, from a black mound in the winter to a verdant forested hillock of tomato seedlings in the early summer!

Although *S. pimpinellifolium* – which I have growing wild in my garden – is rather nice to eat, one cannot say the same for the fruits of most other species of wild tomato. Many academics have scratched their heads over just why the indigenous populations appeared neither to eat or attempt to domesticate them. Although not highly toxic, they can give you a bad stomachache and their sourness makes them an unattractive food to forage. But maybe the reason is obvious: they just didn't taste great. Peru, which lies in the middle of the native homeland of the tomato, is one of the most important regions on Earth for the diversity of its food crops. There is plenty of archaeological evidence of other members of the Solanaceae family being cultivated there, including potatoes and chillies, yet not a single mention of the tomato is to be found, and even today they are not an intrinsic part of native South American cuisine.

The Road to Mexico

At this point in the story of the evolution of the cultivated tomato, the variety *S. lycopersicum* var. *cerasiforme* (a named variety) comes into the picture and has a crucial role to play. Its evolution was the result of a first step in the early domestication process in northern Peru and possibly southern Ecuador and a subsequent 'improvement' in its domestication in Mesoamerica, which resulted in larger fruited varieties. It was the Mayans of this region who were the first great breeders of tomatoes, creating a diversity of colours, shapes and sizes by exploiting *S. pimpinellifolium* and *S. cerasiforme*, which are both interfertile, meaning they can freely cross with each other.[2] Botanist Charles Rick (1915–2002), tomato hunting in the Americas after World War II, was able to collect a huge number of wild tomato accessions (examples of the same variety which were growing in different locations and demonstrated different traits). Today, the largest collection of these accessions, numbering 1,500, is held at the Tomato Genetic Resources Centre in Davis, California, which was established by Rick in the 1970s. It is a vital institution for achieving a greater understanding of the evolution of the cultivated tomato and an invaluable genetic resource for further research into new cultivars.

I like to put myself into the minds of the native people of the first and second centuries BCE. There is no evidence that tomatoes were brought north from Peru and Ecuador overland into Mesoamerica by migrating tribes (if they had carried seed with them, they would have left a plant trail along the route). When I suggested to the biologist and wild tomato expert Thomas Städler that the native population of the northern Andes might have travelled by boat to Mesoamerica, he thought I was having an original idea! The first settlers in South America made their way from Asia between 30,000 and 16,000 years ago, before the melting ice caps cut off the land bridge (now the Bering Strait), making the journey only possible by boat. If people had not migrated north through the inhospitable and difficult terrain between South and Central America, then how had the tomato arrived in Mexico, that part of Mesoamerica where it underwent the greatest change in fruit size and shape? However they came, travellers would have carried useful seeds with them and maybe some survived in the bottom of their boats. The archaeological evidence of domestication is short on detail but why should that stop me from imagining a colourful scenario?

The Spanish called the red fruit *tomate*, which is a bastardisation of *tomatl*, its name in Nahuatl, the language of the Aztecs who were cultivating it at least 700 years before the Spanish showed up. In fact, many Mexican Indian cultures, including those of the Mayans and the Zapotecs, grew the tomato and gave it their own unique names. Long before the arrival of the Spanish, farmers selected for fruit colour: tangerine, yellow, pink, white.[3] They also selected mutations of *S. cersiforme* with two chambers (bi-locular) for those that produced multi-locular fruits, which led to an increase in size. Mutations that gave pear and plum shapes were also considered highly desirable, thus further increasing the diversity of varieties. Mexico continues to be home to the largest number of different varieties of tomato anywhere on Earth. However, *tomatl* was a general name used by the Aztecs for tomatoes and their close relative, the tomatillo.

A Place in Everyone's Heart

The botanist J.A. Jenkins[4] gives a long list of Aztec names for different types of tomato, including star tomato, deer tomato, eye of a deer, little tomato, sand tomato, field tomato, red tomato, plum, gourd, peach, kidney and ribbed, which underlines the skill of these native growers in creating different varieties. Jenkins also makes a telling observation about indigenous cuisine as a result of a visit he made in the mid-1940s. Mexicans preferred to eat the local tomatoes, including wild varieties which were both foraged and grown in the Veracruz and Jalisco regions southwest and northeast of Mexico City. These tomatoes were never exported. Reading Jenkins' accounts of how many Mexicans considered small-fruited foraged tomatoes – sold in great bunches in markets – tastier than the cultivated ones brought a lump to my throat. Until the 19th century, every type of tomato known to cultivation, as well as naturally occurring hybrids unknown anywhere else, were to be found in Mexico.

At the time that Jenkins was plant hunting in Mexico, modern tomato breeding was still in its infancy, based around crossing different varieties to create more open-pollinated cultivars with particularly desirable traits. Indigenous cultures around the world maintained a considerable diversity of local varieties and their survival is crucial not only in reinforcing local food culture, but also in providing invaluable genetic material for traditional plant breeding. Today, dominated by international agri-businesses, almost without exception new cultivars are F1 hybrids. However, a cohort of growers continues to propagate and maintain heritage and heirloom varieties for our culinary pleasure.

What's in a Name?

The Italian naturalist Pietro Andrea Matthioli (1501–1577) ascribes the Latin *Mala aurea* and a common name *pomi d'oro*, both meaning 'golden apple', when naming the tomato in 1544. The power of language and culture is such that *pomodoro* remains the Italian word for tomato. According to Matthioli's account, the first tomato to arrive in Europe was yellow; yet there must have been red tomatoes in Italy before 1544 as a red type is also described by him, and one cannot imagine the Spanish conquistadors bringing only one variety home!

Until less than a century ago, determining where the tomato was first domesticated had no consensus. The prevailing view was that domestication started in Peru rather than Mesoamerica because of the name first given to it by botanists and herbalists. Some believed that the tomato was initially brought to Europe from Peru after the Spanish Conquest of 1535, as is evidenced by the names ascribed to it: *Mala peruviana* and *Pomi del Peru* – both meaning 'Peruvian apple'. The botanist Melchiorre Guilandino of Padua (*c*.1520–1589), who like other botanists of the time had little or no understanding of geography and just considered all remote regions as being 'foreign', thought otherwise. In a publication of 1572 he assigns the tomato to the Americas generally rather than a specific region or country and names it *Tomatl*. He also refers to the tomato as coming from Themistitan, which, according to Jenkins, identifies its correct place of origin, as well as the fact that it was introduced to Europe earlier. Themistitan is also spelled Temixtitan which is a corruption of Tenochtitlan, the name that the Aztecs gave to the city we now know as Mexico City. The conquistador, Hernán Cortés (1485–1547), laid siege to Tenochtitlan in

1521 and, no doubt, brought tomato seeds back home with him when he returned to Spain in 1524.[5]

From Golden Apples to Ketchup

Another botanist, Dutchman Rembert Dodoens (1517–1585) describes tomatoes, translated from the Flemish, as 'Amorous apples or Golden apples'. He names tomatoes in various other languages, the English being Apples of Love and Golden Apple, and in French *Pommes d'amours*. This speedy movement of the tomato across Europe was thanks to the enthusiasm of Renaissance herbalists to check out new medicinal plants, although it was considered nothing more than a curiosity. By the seventeenth century, it was only the British who referred to the tomato as the love apple, while the Italians had cemented *pomodoro* into their vocabulary. Another name commonly used was Wolf Peach, which when Latinised becomes *Lycopersicon*. In 1753, Carl Linnaeus (1707–1778) used this name as the species classification for the tomato, *Solanum lycopersicum*. By the end of the eighteenth century, the English-speaking world finally settled on tomato – an anglicisation of Guilandino's name for the fruit possibly, but also from *pomi d'oro* maybe?

The Chinese had a somewhat different linguistic take on describing the fruit. Although there are no definitive references to the first use of tomatoes in China, they entered Chinese cuisine along its coastal regions early in the sixteenth century at the same time as other New World foods were taking over the world. Reminding the Chinese of their own native fruits, they called the tomato the 'Barbarian aubergine' and 'Western Red Persimmon'. Let us not forget too that the word *ketchup*, a name synonymous today with tomato sauce, comes from the southern Chinese Min language word for fish sauce, *kôe-chiap*. Thanks to its introduction to the Malay Peninsular by British sailors in the eighteenth century, the sauce became, in Indonesian, *ketchap*, which is used to describe any sweet or savoury sauce. The British liked this spicy sauce – calling it *katchup*[6] – to liven up their bland cuisine, adding other fishy ingredients such as anchovies and oysters as well as walnuts and mushrooms; in many ways more akin to that classic British condiment Worcestershire sauce. Apparently, Jane Austen liked the sauce made with mushrooms. In 1812, the American horticulturalist James Meade was inspired to make the sauce with tomatoes, brandy and spices, all sweetened with sugar. The rest, they say, is history!

From Fear to Fashion

For the best part of a hundred years after it first arrived in Europe the tomato was admired by the Italians as a decorative plant – something to adorn the table, but certainly not to be eaten. Its association with both mandrake and deadly nightshade assured this reputation. The 'love apple' or *pomme d'amour* moniker referred to the belief of some that the tomato was an aphrodisiac – the soft, squashy flesh was synonymous with the vulva. Physicians and herbalists of the Renaissance and later periods considered aphrodisiac plants to be both hot and moist. These traits made them overly nutritious for the average human – leading to 'venereal' impulses, apparently. The tomato may have been moist, but it was thought of as cold too, which meant that, rather than making the eater amorous, it made them ill.[7]

Despite this rather inauspicious start, by the end of the sixteenth century some people were eating tomatoes, more out of curiosity than desire. It was, however, thought good enough for peasants and workers, with their strong stomachs, who were hot and sweaty most of the time anyway and so immune to its dangers. But what was the point of the tomato, really? It didn't have any fortifying or nourishing qualities. Generally, foreign food was frowned upon as being inferior to what we already had. This was a commonly held belief across Europe in the sixteenth and seventeenth centuries, and further underlined by a dislike of vegetables in general which, if not completely avoided, were only ever eaten in limited quantities. Physicians of those days had nothing good to say about the love apple.

Low-Growing Status

The tomato had other problems too. It was not easy to grow in the rather dull, wet and cooler northern European climate and, even in countries like Italy where it grew with glee, its trailing habit did it no favours. There was generally a prejudice against low-growing crops, which were given an equally low health status. Today, those same traits of indeterminate growth – where the lead shoot just keeps growing until something like frost or a herbivore stops it – are exploited by growers who train the vines vertically along strings and canes, ease of harvesting and bigger yields in smaller spaces being key.

It wasn't until the middle of the seventeenth century that the tomato's qualities as a culinary delight became better appreciated and it started to lose its sinister reputation. The English naturalist and gardener John Ray (1627–1705), who documented his travels to Italy during the 1660s, wrote that the tomato was cooked 'with marrows, pepper, salt and oil'.

In Italy the tomato was being used as a condiment. The first tomato recipes are found in the work of the seventeenth-century Italian chef Antonio Latini, who described three ways of using tomatoes, all in the Spanish style – including the first recipe for an Italian form of ketchup. Spain ruled over much of Italy at this time, which perhaps explains the culinary techniques.

Being Self-Contained Has Its Advantages

Until the end of the eighteenth century, English seed merchants listed tomatoes as love apples. In 1760, John Webb sold a solitary love apple with no description. In his 1780 Catalogues for Seeds and Plants for the Kitchen Garden, J. Gordon of Fenchurch Street London listed two varieties for sale: red-skinned and yellow-skinned. Things were better on the continent, with a diverse range of local varieties of all shapes and sizes becoming part of regional cuisine. This increase came about through natural and accidental crossing. Tomatoes are generally self-fertile, meaning they are able to pollinate themselves, but they can still cross with other varieties if conditions are right. There are three distinct types of tomato flower: ones with protruding (exserted) stigma; those with inserted stigma contained within the flower; and, finally, a double-flowered or marigold type which is synonymous with certain very old and large beefsteak varieties. All are self-fertile with hollow, pollen-containing anthers.[8] In the case of tomatoes with inserted stigma, the pollen is released when the flower vibrates (this is the reason growers like me give their flowering tomato plants a good shake morning and evening). Commercial growers have machines to do this vibrating instead (although there are still some who introduce bees into their glasshouses to do the job). In nature the clever bee hangs onto the bottom of the flower, vigorously vibrating its body which causes pollen to fall. This is why tomatoes will occasionally cross because pollen accidentally attaches to the style of another flower. This was the primary cause of new varieties, effectively breeding by accident. Only later in the nineteenth century did breeders become more systematic in their efforts. Today, almost all commercially available varieties have inserted stigma. There are lots of heritage and heirloom varieties that still have exserted stigma, which means a seed saver like me must be careful to isolate such varieties from each other, to ensure no crossing. In fact, it is really very difficult to get tomatoes with inserted stigma to cross because the pollen is usually viable before the flower is fully developed. Hence, a good shake tends to initiate early pollination and is why these types of tomato are such an easy crop from which to save seeds.

An American Journey

If things were tough for the tomato's reputation in Europe, they were equally bad in the USA. There is some evidence that tomatoes came to the southern states of the USA early in the seventeenth century when Spanish settlers moved into the Carolinas from the Caribbean. They were definitely under cultivation during the eighteenth century, although early settlers had no time for them, having brought their European prejudices with them as they migrated across the Atlantic. The tomato does not appear to have travelled overland from Mexico to be cultivated by the indigenous tribes of North America either. Thomas Jefferson in his *Notes on the State of Virginia*, published in 1781, mentions the tomato as an edible crop. In 1809 Jefferson, or rather his slaves, started to grow them at his New York State home in Monticello. But the northern states took a lot of persuading to eat the things – again due to the fact that the climate was less than ideal for their cultivation but also because the predominantly northern European immigrants either knew nothing of the tomato or feared it.

During the 1820s Americans finally came around to eating tomatoes and seed merchants listed one or two varieties, which were similar to those available in the U.K. nearly half a century earlier. I particularly like the colourful and fictitious story of a certain Colonel Robert Gibbon, who in the autumn of 1820 stood on the steps of the courthouse in Salem, New Jersey, and, in front of a large crowd and against the advice of his physician who suggested such an act would end in appendicitis, ate a basketful of tomatoes. The tale was widely believed and did presage a change of fortune in the tomato's journey to the very heart of American cuisine. Suddenly, this humble fruit became the panacea for all ills. Extract of tomato became a cure-all, consumed by anyone with the slightest ailment. Alongside snake oil, in your average American quack's bag could be found many variants of tomato extract, most of which contained no tomato. By the middle of the nineteenth century, thanks to its supposed curative qualities, the USA was going tomato mad, as was practically the entire world.

Tomatoes cultivated in the USA tended to be larger and flatter than those from Europe. By the 1850s American breeders started to select for uniformity as well as for size and colour. Probably the first successful cultivar, the result of more than 20 years of systematic crossbreeding, was the Trophy tomato bred by Dr Hand of Baltimore County,

Maryland. By removing the anthers from one type and brushing them on the style of another, there is some evidence that Trophy was the result of crossing two common American varieties, 'Large Red' and 'Early Red Smooth'. Starting in 1843 and, after some trial and error, Dr Hand produced an apple-sized fruit with a smooth skin that was glossy and an even red colour.[9]

Whether named by Dr Hand or his savvy future business associate, Colonel George E. Waring of Rhode Island, the Trophy tomato became a sensation.[10] Introduced to an obsessed public in 1870, Waring had the front to sell a packet of 20 seeds for $5, which was more than the cost of a gold ring. He offered a prize of $100 for the heaviest tomato. Needless to say, gardeners across the USA bought the seed and Waring made a fortune. The Trophy tomato was a bestseller in the USA for decades, as well as being the progenitor of numerous other cultivars which remain part of a breeding line that continues today.

A Case of Local Identity

The nineteenth century was an extraordinary period for plant breeders. By the last quarter of the century seed sellers would have at least a dozen varieties of tomato available. In 1885, Vilmorin-Andrieux in France had 15, including the King Humbert, which was named after the Italian king Umberto I (1844–1900) whose consort Margherita inspired the famous pizza. King Humbert was trialled in 1887 by the Royal Horticultural Society, who described it as 'An extraordinary cropper...'. Hundreds of locally named varieties were bred, many of which are still around. But it was the Italians and Spanish who truly embraced tomato localism.

It would appear that every part of Italy has its own tomato: Pantano of Rome, Costoluto of Parma, Cuor di Bue of Liguria, Costoluto Fiorentino from Rome. For pizza lovers at least, one of the most important tomatoes in the world is the San Marzano. Revered by Italians – well, Neapolitans at least – this tomato has *Denominazione d'Origine Protetta* (D.O.P.) status, meaning only tomatoes grown in this region on small farms in the foothills of Mount Vesuvius can be sold as San Marzano. Needless to say, thousands of tons of tinned tomatoes are sold all over the world under that name, but they are not grown beneath Vesuvius! Another case of stolen identity perhaps?

The Spanish have at least 150 named traditional cultivars. Some, like the Raf tomato which I came across in the Níjar region of Almeria in southern Spain, is happiest growing in slightly saline conditions as a winter crop. Bred in France and first commercially available in the early 1960s, Raf (which in French is the initials for Resistant to Fusarium) is the result of a cross between Marmande (a large-ribbed type) and a commercial cultivar, Red Global.

Sadly, the fate of these unique tomatoes is in the balance. Farmers no longer focus on growing delicious but lower-yielding local types. Instead, they prefer the higher yields and profits of the latest F1 hybrids (from which the farmer cannot save seed, as they don't grow back the same). However, plant breeders still need access to as wide a genetic resource as possible, so tomatoes like the Raf continue to be maintained by its French breeder and can occasionally be found in trendy veg shops across Europe during the winter months. Fortunately, as it is an open-pollinated variety, I am able to save seed and continue to grow and savour this wonderful tomato.

The Spanish have been quite expert in developing cultivars that can be harvested in the autumn and ripened through the winter. On one of my most memorable seed-hunting adventures I was celebrating a birthday on a walking holiday in the Garrotxa region of northern Catalonia and visited fellow seed saver Jesus Vargas. It was early spring, and I was able to enjoy a typical Catalan way of eating tomatoes. As in many other regions of eastern Spain, the pulp is spread on bread like butter. *Pa amb tomàquet* (Catalan for 'bread with tomatoes') is often eaten for breakfast. It is traditionally made using a special thick-skinned tomato that is stored and ripens through the winter months. Jesus gave me two such native varieties, *Lagosterz Tomate de Penjar* (Hanging Lobster Tomato) and *Piel de Melocotón* (Peach Skin). Hung up as ripening trusses in a cool and airy shed, the tomatoes can keep until early spring. Although not sweet, their flavour is as enjoyable as their names. Although there are countless different varieties of tomatoes bred throughout northern Europe, we really need to head back across the Atlantic to track down those that not only have great flavours, but also great names.

American growers started a tomato breeding frenzy in the latter part of the nineteenth century and today a handful of breeders continue this tradition. Checking boxes of tomato seed in my fridge, I am reminded of pleasures past and those to come. Wonderful American heritage tomatoes include Salt Spring Sunrise, Speckled Roman, Bolivian Orange Cherry, Broad Ripple Yellow Currant and Brandywine, so named by the teetotal Amish farmers after their local river. Some of the finest tomatoes I have ever grown are Cherokee Purple, Lilac Giant and Amish Yellow. The list goes on and on. Books abound that catalogue many hundreds of varieties from all around the world that, as a result of the globalisation of our food, are now grown everywhere.

Breeding Gets Serious

With the arrival of the tomato in Europe, growers became very good at selecting lots of colourful and delicious traits, which was just fine when the crops were mostly consumed locally. Selection was on a single plant basis, saving seeds from small numbers. Because most tomatoes are interfertile, seeds from the parent plant produced identical offspring. Thus, seed was saved within families and communities and has come under the collective description of heirlooms. Until commercial plant breeding took off late in the nineteenth century, the vast majority of tomatoes were heirlooms. Breeders began exploiting the infrequent crossing of open-pollinated varieties which produced different types. This was not a problem until, in the latter half of the century, everyone decided they loved tomatoes so much that demand exceeded supply.

Suddenly, the world needed tomatoes that could be transported from one end of a country to the other. Tomatoes that could support a burgeoning canning industry, thanks to demand first generated during the American Civil War (1861–1865), needed to be uniform and ripen more evenly, thus making harvesting more efficient. A glut of tomatoes produced after the war was seen as an opportunity by some growers to form a 'Tomato Trust', effectively to monopolise and corner the canning industry. However, a boycott by tomato middlemen and other suppliers put paid to that. Their story, however, illustrates just how revolutionary the change of attitude in America towards the tomato had become.

From Open-Pollinated to F1

Any gardener will have come across F1 hybrid tomatoes when perusing seed catalogues: the first was Single Cross, released in 1946 and bred for appearance, yield and shelf life. Only in the last few years has taste been considered a worthwhile trait. Today, supermarket shelves are filled with colourful additions alleging nutritional quality and flavour. These latest cultivars have higher levels of the naturally occurring antioxidant lycopene, which gives tomatoes their red colour and is more easily absorbed by the body when cooked.

With just a handful of tomato breeders globally producing hybrid seed, competition is brutal. Maintaining the genetic resources to create new hybrids is very costly and breeders need to constantly come up with new cultivars. There is much innovation in their secretive and protective business. The value of the intellectual property of F1 hybrids demands that they must deliver volume and eating quality on a massive scale to justify their expense. Consequently, the average life of a commercial tomato is just five years. The annual global tomato seed market is worth nearly a billion dollars.[11]

Appearance is one thing, but for me, and surely for most tomato lovers, the thing that matters most is flavour. How a tomato tastes is down to a series of interactions between sugars, acid and several volatile compounds. It is possible to make some predictions on the likely flavour scientifically by measuring the acidity and amount of solid content in the fruit. Some success in assuring a certain flavour with the right balance of sugar and acid occurred with research in employing genetic modification (GM).[12] However, breeding for flavour has proven rather difficult and, as anyone who grows tomatoes knows, temperature and light levels are fundamental contributors to our culinary enjoyment.

Breeders became very excited about the potential of using GM to develop new lines of tomato. The first and only GM tomato to be eaten fresh was Flavr Savr (no spelling mistake I assure you), which was brought to market by Californian biotech company Calgene in 1994. It was bred specifically for shelf life. Generally derided as tasteless despite claims to the contrary, it was withdrawn from sale by 1997. It was too expensive to propagate, but the main reason for its demise was consumer resistance to eating 'Frankenstein food'. Calgene did not thrive either, being acquired by Monsanto some years later. Zeneca bred a tomato similar to Flavr Savr for the tomato purée market. It was cheaper to produce than conventional crops and, in the final years of the twentieth century, nearly two million tins of the stuff were sold as an 'own-brand' in some British supermarkets. Clearly labelled as GM, it was a bestseller for a short time. However, by 1998 sales were falling, again due to negative public opinion. The supermarkets in question pledged not to use GM ingredients in any of their 'own-label' brands and that pretty much marked the end of the GM story with tomatoes.

The first gene-edited tomato was approved for sale in Japan at the end of 2020. Seedlings have been enthusiastically taken up by commercial growers and seeds for amateur gardeners go on sale in 2022. Known as the Sicilian Rouge High GABA tomato, it has higher levels of an amino acid that lowers blood pressure, so is good for your health apparently. Just how successful it proves to be only time will tell.

Today breeders use their understanding of the genetics of tomatoes with desirable traits alongside traditional cross-breeding to create new cultivars. Some come to market using the long-tried and understood techniques of conventional breeding – especially in the USA, where traditional breeding practice is performed by many skilled small-scale operators. Yet it is the F1 hybrids that dominate the world market.

Contrary to the opinion of some in the world of sustainable agriculture, not everything new in plant breeding is to be condemned. F1 hybrid tomatoes are much better on the flavour front these days; they have improved disease resistance and greater yields, and along with lots of shapes, sizes and colours, more people than ever are eating them. In fact, tomatoes are the number one vegetable grown in the world; 16 per cent of all cultivation is down to the humble tomato.

....

It is spring at home. The sun is shining. The half-dozen or so varieties I sowed a few weeks ago are demanding my attention. They need transplanting from trays in the propagator into individual pots. Already the immature leaves exude that distinctive tomato aroma when I touch them. I am filled with optimism and hope they will grow happily and well, and already I am impatient for that moment when I can pluck the first ripe fruit from a vine, hopefully in early June. Maybe there will have been an accidental crossing from a previous year and I will be presented with something new and wonderful this summer: Adam's tomato. Has a nice ring to it.

A Very Uncommon Bean

Beans, beans are good for your heart.
The more you eat, the more you fart.
The more you fart, the better you feel,
So, eat your beans with every meal!

Anon

It was September 1981, and the shrouded peaks of Italy's eastern Dolomites' vertiginous, shark-tooth summits were lost in autumn mist. This was the view that presented itself to me as I sat in the municipal hall alongside the good citizens of the pretty town of Belluno. I was there making a film about a remarkable British war hero and adventurer, Bill Tilman (1898–1977). The occasion was a banquet to celebrate the naming of a street after him. The event fortuitously coincided with the completion of the harvest of a crop unique to this corner of Italy. Being a very Italian affair, the food was simply delicious and the atmosphere boisterous. It was here that I was introduced to the bounty of the harvest, a bean that has been a faithful friend ever since, *Fagioli di Lamon* (Lamon bean).

I remember a lovely dish of pasta with fat, brown beans in a delicious vegetable sauce. On enquiring as to its provenance, the lady I was sitting next to started to expound on the bean's uniqueness and magnificence. There are four types of *Fagioli di Lamon* grown on the nearby plateau after which the bean gets its name. The variety I had enjoyed was *Spagnol*. Today, this bean and its sisters, *Spagnolit*, *Calonega* and *Canalino*, have special status under EU law and, since 1993, are grown and marketed through an organisation known as *Consorzio per la Tutela del Fagiolo di Lamon della Vallata Bellunese IGP* (Consortium for the Protection of the Lamon Bean from the Belluno Valley PGI).[1] The cultural and social significance of this bean – with its beautiful, red-and-white-striped pods and speckled red and white beans – was rather lost on me at the time. I had the temerity to suggest that it was just like the borlotto bean I already grew. I thought the blossoming friendship with my fellow diner would end right there. I had insulted her bean. My apology was accepted and a wish to grow the Lamon bean instead of the pale, vulgar imitation I so foolishly had in my garden was granted. From the kitchen emerged the chef with a small bag of beans that I have been saving and growing ever since. Little did I know at the time that this encounter with a bean would be repeated many, many times over the next 40 years.

The Borlotto of Belluno

The *Fagioli di Lamon* is a type of borlotto bean, a variety of a group known as cranberry. It originated in Colombia, where it is known as the *cargamanto* bean, and was brought to Italy, if you will believe the local narrative, by Hernán Cortés (1485–1547), who had acquired it from the Aztecs when he conquered Mexico in 1521. Cortés, so the story goes, gave the bean to Charles V who was both King of Spain and much of Italy at the time. He passed it on to Pope Clement VII who then gave it to the Renaissance humanist Piero Valeriano

(1477–1558) in 1532. So revered was it by Valeriano that he wrote about it in truly florid terms (although there remains some doubt about whether these are his words):

'Father Clement himself gave me a gift from far away.
And giving it he said, "You shall enrich your
  homeland hills.
With a new fruit, you shall gladden the fields of Belluno.
Therefore I, when I returned to my homeland,
  sowed these given seeds
Not in fields, however, nor entrusted them to gardens,
  but rather
To adorn my dwelling with earthen pots, and my
  windows with saucers.
Surely hoping for some very small crop from them;
  and, behold!
First there arose a prodigious forest of leaves,
Everywhere flowered countless violet blossoms
  throughout every
Tendril, and all of them full of pregnant pods.'[2]

A Cinderella Bean

This much-loved vegetable is just one of countless varieties of *Phaseolus vulgaris*, the common bean, otherwise known in the U.K. as the French bean. We have Christopher Columbus to thank for introducing this bean to Europe too. However, unlike its relative, the lima bean, which he also brought back with him from the West Indies, the common bean entered European food culture under the wire. It apparently joined the pantheon of 'peasant food', crops that were considered only fit for the poor. Farmers probably assumed the common bean to be much like all the other similar-looking ones being consumed across Europe at that time, including the cowpea (which is native to Africa) and fava beans, without giving any thought as to its origin or even name. Common bean and cowpea pods look much alike. But before continuing with the story of how this, in my view very uncommon, bean became a global food phenomenon, let's first look at its origins.

The wild parents of the common bean grow in an area that stretches from northern Mexico to Argentina which, for this purpose, can be divided into two broad regions: Mesoamerica (including Mexico) in the north and the Andes in the south.[3] Wild common beans endemic to the two regions are genetically distinct and, because of their morphology, have given rise to a huge number of landraces that are regionally specific. This has resulted in reproductive isolation which has remained over millennia. Their genetic divergence suggests that these wild beans evolved from a common ancestor found in Ecuador and northern Peru into two sub-species over an evolutionary timescale of possibly hundreds of thousands, or even millions, of years.[4] The wild common bean found in Mesoamerica is *P. vulgaris* var. *mexicana*, and the one from the Peruvian Andes is *P. vulgaris* var. *aborigineus*. Domestication would have happened independently in the two regions, but which domestication came first and just how long ago? There has been much debate and many papers written on the subject.

The absence of remains of wild common beans in archaeological sites where one might expect to find them, especially in Mexico, suggests that today's wild landraces are the ancestral parents of cultivated landraces which are sympatric, meaning they are closely related to each other and therefore similar to the earliest, but now extinct, wild beans that preceded them. Some have argued that archaeological remains of domesticated common beans are older in South America than Mesoamerica, pointing to an earlier domestication. Now genetic mapping and analysis would seem to have settled the matter. The common bean was first domesticated in Mesoamerica and travelled south, along with its two sisters, maize and squash, into the second area of domestication in the Andes. To keep things simple, plant scientists now refer to all domesticated common beans as a single species, *Phaseolus vulgaris*. As to when Neolithic farmers started to gather beans, archaeological records proving that they were a part of their diet go back at least 9,000 years in Mesoamerican sites and nearly 8,000 years in Andean sites.

A Mixed Bag

Wild common beans are all climbing vines, with twisted pods containing small seeds that shatter when ripe. Dwarfing varieties that grow as bushes would seem to be entirely the result of early human intervention; they have no wild ancestors. Two of the traits that characterise the cultivated common bean are extensive regional diversity and seeds that are larger than those of wild beans. This diversity, which expresses itself in bean colour and size as well as growing habit, is due to the existence of a huge number of landraces.[5]

One of the strongest characteristics of the common bean is that it is cleistogamous. This trait describes plants that are self-pollinating, with non-opening flowers. Once a new variety or landrace becomes established – as a result of a mutation or random hybridisation, for example – it remains stable and distinct, which probably makes the common bean the easiest crop from which to save seed. It is virtually impossible, other than by genetic accident, for crossing to happen, which means that all the saved seed is true to the parent. This trait, as we shall see, has been vital to the success of the common bean as a staple of New World cuisine.

Visit a market in Mexico's highlands and you will be presented with a kaleidoscope of colourful common beans. Piles of dried pulses consisting of maybe half a dozen or more local varieties or landraces of many colours and patterns beg to be bought. They are deliberately cultivated and sold as mixtures, although sometimes individual colours are separated if they can command a higher price. As I recount later, there is one bean in particular that I treasure. However, indigenous Mexican farmers have been employing this method of growing lots of different varieties of beans together for at least the last thousand years. They have much to teach us. Growing multiple varieties of beans together is done for some important reasons. Firstly, these farmers all agree that they get better crops when different beans are grown together and, secondly, because they don't cross-pollinate with each other, each variety's distinct traits are preserved. Germination experiments point to a possible explanation for why mixed planting gives better yields. The time different varieties of beans take to germinate is dependent on two factors: moisture and temperature. Farmers want to sow their crops at the time of the spring rains, which cannot be depended upon. Sometimes the rains are short, so early sowings start growing and then fail. Low temperatures mean slow germination which, in cooler springs with later rains, is of benefit to those landraces that are naturally slower to germinate. The farmers maximise their chance of a crop surviving with this method of cultivation. Although yields from heavier cropping varieties may be less as a result, at least they have a crop to trade and don't starve. When it comes to harvesting, the pods of vining (climbing) common beans ripen at different rates, so are collected over several weeks. This is a benefit for farmers' time and the availability of additional labour. The traditional method of sowing lots of different varieties together is a feature of highland farming. The higher soil temperatures found in lowland farms enables more uniform and faster germination, so farmers focus on single, higher-yielding varieties.

We are often led to believe that the process of domestication came about as the result of farmers selecting from observed changes in a crop's morphology, thus creating new and better varieties. However, wandering in the markets or among fields of growing beans today, we see the outcome of a different process of domestication. The wealth of diversity – all those shapes, sizes and colours – is the result of selection from many wild landraces that are naturally adapted to the farmer's preferences. The accidental gene flow between cultivated landraces and their neighbouring indigenous wild relatives would have, over millenia, created beans with traits the farmer didn't want, such as twisted pods, hard seed coats and small beans. These would have been eliminated over the time when domestication occurred.

Sometimes you can find single varieties of beans in Mexico's highland markets. One such, much prized by me for its fine eating quality and heavy crop, is a small, semi-dwarfing type called Black Delgado, which is native to the Oaxaca region of southern Mexico. It is one of just three landraces that are found growing in regions which overlap with archaeological sites that have yielded well-preserved bean remains. Known locally as *Frijol Delgado*, it continues to be widely grown and is an important part of subsistence peasant agriculture. Many academics have concluded that it is an indigenous landrace that has been in cultivation for millennia. Growing vegetables that are the direct descendants of those cultivated by the world's first farmers reinforces the sense of continuity that connects me directly with my distant ancestors. Oh yes, and Black Delgado tastes good too – especially in a spicy bean stew.

At the same time as native beans from the eastern Mediterranean were deeply embedded in Egyptian culture, the cultivation of a hugely diverse number of varieties of common bean, both climbing and bush types, were cementing their own place in Mesoamerican and Andean cultures. There is little doubt now that the very first Neolithic farmers in South America were domesticating their beans at the same time as their distant relations from the Fertile Crescent were domesticating fava beans. Indeed, there is now a lively debate among academics about whether agriculture might have started first in the New World – and certainly the skills of its indigenous farmers were equal if not superior to those on the other side of the world.

Common beans, however, were literally intertwined with two other crops that made up the diet of the indigenous people in the New World: maize and squash. Although there is no reason to think that earlier farmers were not growing these three crops sooner, the archaeological record suggests that they were only being widely cultivated together in Mesoamerica 3,500 years ago, and much later in North America. Named the 'Three Sisters' by native North Americans because they could only thrive when grown together, I like to employ this method of cultivation myself. The beans, which fix nitrogen from the air on their root nodules, like to climb up the nitrogen-hungry maize plants, while underneath squash plants scramble, smothering weeds and reducing evaporation – very important for farmers such as the Hopi subsisting in the arid Arizona desert. This form of multi-cropping – growing different types of vegetable in the same place – is highly beneficial to soil health and fertility. It is the polar opposite of the modern farming practice of monoculture, which calls for high inputs of fertiliser and chemicals as a mitigation against limited or no crop rotation. It is now widely agreed that mixed planting improves soil health and fertility and results in greater carbon sequestration.

An Invisible Import

At the time of the arrival into Europe of the common bean, all species were a part of *cucina povera* (the food of the poor), so never appeared on the dining tables of the rich and cultured. Lumped together, all beans were called *fagiuolo* in Italian and *fasoulia* in Arabic, from the Latin *phaseolus* via the Greek for cowpea. In 1542, the German botanist Leonhart Fuchs (1501–1566) doesn't mention the bean coming from the New World. Incorrectly, he says it is synonymous with the cowpea but names it *Smilax hortensis*, which is not a legume, being of a different genus and often known as sarsaparilla, a key ingredient of a very American soft drink. It took another decade before the penny finally dropped and botanists recognised that the common bean was entirely different to the other beans being consumed by peasants. The Flemish botanist Rembert Dodoens (1517–1585), who is often considered to be the Father of Botany, goes to considerable lengths to point this out in his famous herbal *De frugum historia*, published in 1552.

Although botanists at this time identified the common bean as a new species being cultivated in Europe, they didn't know where it came from. The first person to identify the common bean's native homeland was physician and botanist Castore Durante (1529–1590). He lists the medicinal qualities of a new bean from the 'Indies' – the New World – in his book on health remedies for the family, *Il Tesoro della Sanità* (Treasury of Health), published in 1586. In it he describes the bean's rather exciting quality, for men at least – especially regarding beans coloured red – of producing more sperm as well as exciting coitus. Are there any vegetables unable to do wonders for one's sex life?[6]

The first English description of the common bean appears in the seminal work *The Herball or Generall Historie of Plantes* (published in 1597) by the botanist John Gerard (1545–1612). In it he describes two types of climbing 'Kidney Beane': a white one that he calls *Phaseolus alba* and a black one, *Phaseolus niger*. He classifies them as 'garden Smilax' (because of their twining habit) and describes two dwarfing types: Red Kidney Beane (*Smilax hortensis rubra*) and the Pale Yellow Kidney Beane (*Smilax hortensis flaus*). There then follows a description of more varieties, all with names that illustrate the undisciplined world of plant classification at the time: Kidney Beane of Brasile, *P. Brazilianus*; Purging Beane of America, *P. Americi Purgantes*; and the Partly Coloured Kidney Beane of Egypt, *P. Ægyptiacus*. His descriptions include many seed shapes and colours.[7] There is no doubt that by the end of the sixteenth century the common bean in many forms was widely known. The Egyptian bean described by Gerard can be found under cultivation in England and the USA, going by the name of pea bean. Its beautiful, bi-coloured (deep purple and white), round seeds can be dried – young pods also make good eating – just as they were in England over 400 years ago, when it was eaten mostly as a fresh green bean but also pickled and salted to provide much-needed variety in a still very dull winter diet. Like many common beans, the dried seeds were also a useful and nutritious addition to winter soups and stews.

The curiosity displayed by botanists throughout the second half of the sixteenth century was driven in part by the impact of printing. This revolutionary new technology enabled the dissemination of knowledge to be shared widely. It was also due to the arrival of a huge number of previously unknown plant species, which had not been identified or classified by ancient minds. This forced botanists and herbalist into a complete taxonomic re-think. Until they started correctly describing these new members of the bean family, it is only possible to surmise that both bush and climbing common beans were being grown in southern Europe. The fact that Spanish and Portuguese commentators didn't mention them as being distinct when Columbus returned with some in 1494 suggests that he too presumed they were related to cowpeas.

An Italian Love Affair

It was the Italians who first took the common bean to their hearts. *Fagioli di Lamon* may have been the earliest named Italian variety, but it is the Tuscans who truly fell in love with the common bean and proudly continue to call themselves *mangiafagioli* (bean eaters). How well I remember sitting in the kitchen of a country hotel, set among the vineyards and smallholdings outside Montepulciano where I was making a TV cookery series. I was being instructed by chef Giancarlo Caldesi on the finer points of preparing *ribollita*, a soup that includes any and all available beans, as well as another Italian classic, cavolo nero kale (of which you should know I am no fan), and leftover bread, all baked in the oven. This dish is a fine example of *cucina povera*. Peasant farmers, having to subsist on reliable and nourishing vegetables, created a range of simple and wonderful bean dishes which now populate Italian cookery books for those of us who want to find new ways to enjoy them.

Probably the most revered of all Tuscan beans is *Zolfino*, so named because of its sulphurous colour. Today it too has Protected Geographical Indication (PGI) status and is grown by a handful of farmers in its homeland, Pratomagno, a region consisting of just three small villages nestled in the rolling hills southeast of Florence. Not only Tuscans love them. They are considered by many gastronomes as the finest of all shelled beans, with a thin skin which melts in the mouth like *l'Ostia* – 'sacramental bread'. Today, Italian seed catalogues are stuffed with many distinct types of borlotto beans, all associated with a region of northern Italy. White cannellini beans, which are common in central and southern Italy, are one of the commonest white beans, known also as flagelot in France.

It wasn't just shelling beans that the Italians developed over the following centuries. By the end of the seventeenth century, fresh beans were considered fit to be served at the tables of the wealthy. They came in various colours: green, yellow and purple. Today we know them by the catch-all name of French beans. Varieties were named after the villages or towns in which they had become intrinsic to local food culture and cuisine. Some beans have names that beg to be eaten: Meraviglia di Venezia, Trionfo Violetto, Stortino di Trento.

Fresh beans may have been more popular among the wealthy because they were less likely to cause flatulence – a curse for many bean eaters and those close to them. The build-up of gas in the alimentary canal is thanks to a particular complex carbohydrate produced by beans called oligosaccharide. These are large molecules that, because we don't produce the enzyme that breaks them down into simpler sugars, which can be absorbed through the small intestine, reach the large intestine intact. Here bacteria are able to break them down instead. This causes fermentation, which produces the gases that cause flatulence.*

The Italians may have been bean trailblazers, but it didn't take long for this wonderful pulse to find its way into the hearts and stomachs of diners across Europe. It is most likely that French peasants, like other European farmers, did not differentiate between the common bean and the cowpea. When the French finally decided to make it part of their national cuisine, they gave it a new name, haricot, which first appears in French literature during the seventeenth century. It is thought that the name is taken from the Aztec Nahuatl word for the common bean, *ayacótl*.[8] The changing social trends that occurred first in the sixteenth century, when botanists realised that the 'ancients' didn't know everything about the plant kingdom, were followed towards the end of the seventeenth century by a romantic curiosity about rustic life, resulting in a craze for gardening, creating idyllic landscapes and growing exotic veggies. By then the prejudice

---

* Did you know that farting is part of a healthy digestion and something we all do at least fourteen times a day, regardless of how many beans we eat? People who claim never to 'pass wind' are being economical with the truth.

against all beans was overcome and both fresh and drying types became more widely accepted as garden crops. At first, however, it was the flowers of the common bean that were most prized. Red beans were considered the best to grow for their fresh pods, and they seemed to go down well among more discerning Parisian diners.

A Bean Full of History

Native Americans were growing beans when they first migrated into the southern states of today's USA at least 3,500 years ago. Many traditional varieties are a regular sight in my veg plot, busily producing pounds of both fresh and shelling beans. Their names are evocative of tribes and regions: one of my favourites is a beautiful climbing bean whose green pods turn a deep purple as they mature. Eaten fresh they taste great, but their real value is as a dry bean. The small, black seeds were vital to the survival of the Cherokee people who took them with them when they were forcibly relocated from their homeland in the Carolinas to a reserve near Oklahoma City during the brutal winter of 1838. One of the greatest acts of betrayal by the US government, some estimate that as many as 5,000 died on the journey, the route of which is known by the Cherokee as The Trail of Tears. After this act of ethnic cleansing, they named their bean Cherokee Trail of Tears.

America has countless named heirloom and heritage varieties, although many are identical apart from the name! Beans have been baked by the Native Americans for thousands of years, but it was as a subsistence meal on the wagon trains heading west in the nineteenth century that 'baked beans' became an icon of US cuisine. Any available dry bean, sometimes bartered with the local native population, would be slowly cooked while the wagon trains were corralled at night and then eaten for breakfast and dinner the following day. After a hard day rolling across the plains the last thing people wanted to do was cook. Light a fire, warm up the beans and prepare more to cook another meal through the night in a pot known as a Dutch oven which would be buried over hot charcoal. Baked beans were nutritious and cheap. More than simply food for the poor, the dish was synonymous with the ideals of frugality and industriousness, traits that were considered both godly but also as the path to prosperity. This dish, which often included fatty pork or bacon, is a far cry from that with the same name that became Britain's favourite tinned food.

A Star Is Born

Founded by Henry John Heinz in 1869 in Sharpsburg, Pennsylvania, the company which bears his name started out by making horseradish sauce. However, reliant for profits on such a product, one should not be surprised that he quite quickly went bust. Undaunted, he set up a new company with his brother and a cousin to produce a more marketable product – tomato ketchup. He subsequently bought out his family partners and came up with the familiar brand of '57' varieties, just because he liked the number.

Navy beans, so named in America because they were widely used to feed sailors from the start of the nineteenth century, are similar to the French haricot – white, slightly oval and small. They are grown as a bush bean and, thanks to their trait of uniform ripening, are easy to harvest mechanically. Due in part to the impact of imperialism by European powers in the nineteenth century and the need to feed their armies, navy beans were cultivated throughout the Far East and South-East Asia. The US military were so concerned about food security during the war in the Pacific that from 1942 they had the Australians grow sufficient to supply their troops stationed in Queensland, where it was known as The Yankee Bean. Today the navy bean is a global commodity, grown in Brazil, India, China, Myanmar, Mexico and the USA, making up a major proportion of the 18 million tons of common beans traded annually. This was the bean used by Heinz in his most successful tinned food, baked beans, which he started to export to Britain in 1886, selling it in London's Fortnum and Mason as a luxury item. Twenty years later everyone, rich and poor, seemed to be eating them. They were soon being processed in a factory in Peckham, supplying a market that today is the largest in the world. The British eat one and a half million tins of baked beans a day, equivalent to more than five kilos a year for every man, woman and child. Today the U.K. imports 50,000 tons a year of dry navy beans from the USA to supply a factory in Wigan to manufacture baked beans. During World War II, Heinz baked beans were considered an essential food and its advertising slogan 'Beanz Meanz Heinz', created in 1967, has been voted the best advertising slogan of all time.

Heinz beans aside, the nineteenth and twentieth centuries have been a great time for the common bean. Known in the U.K. as a French bean, no doubt because it was from France that they were first introduced, it was the whole young pod that was most prized. I probably grow a greater number of different types every year than I do any other crop. Perennial favourites are ex-commercial varieties with lovely names that are synonymous with great flavour. The stiletto-thin

Emperor of Russia, a twentieth-century variety from France, is probably my favourite. The English Ryders Top-O'-the-Pole with its huge, long, delicious pods and white haricot seeds, was widely grown in the 1970s, but for reasons I cannot understand went out of fashion.

The last one hundred years has seen a revolution in bean breeding. Traditional varieties remain a bedrock of seed catalogues, with many modern open-pollinated varieties (from which the grower can save seed), as well as some fine F1 hybrids that have excellent eating qualities, being virtually stringless – a problem with many heritage and heirloom varieties. Although I do grow some modern cultivars, it is on the heirlooms and older varieties that I focus my attention. I am fortunate to be able to enjoy truly ancient examples such as *Fagioli de Lamon*, which has been grown in Italy for nearly 500 years or, like Black Delgado, have been part of Mexican food culture for more than 3,500 years. These beans mean as much to me as my most distant ancestors. They represent a continuum in the story of human civilisation to which I feel powerfully connected. Growing a crop of colourful beans, creating a delicious meal, and then admiring the beautiful seeds evokes memories and emotions redolent of looking through a family photo album.

The Colour of Corn

'Heer, of one grain of maiz, a reed doth spring
That thrice a year five hundred grains doth bring'
Guillaume Du Bartas (1544–1590) –
Divine Weekes and Workes

There are two common types of maize grown by the Native American Hopi people: blue and white. My desire to meet these quiet farmers had been on my mind ever since I first grew a crop of Hopi Blue maize from seed given to me by my cousin Denys. Hopi maize thrives in arid places, so I grew it in a polytunnel and watered it only infrequently during the entire growing season. The plants weren't bothered and grew well, to a height of about two metres. After harvesting the cobs when dry on the stems, I had much fun stripping the beautiful blue seeds to grind them to make polenta. However, for that I needed a small domestic corn-grinder – not something easily obtained in the U.K. Time for a trip to the US. But I had to wait five years before travelling for the first time into the Hopi Reservation in Arizona, on a quest not only for a corn-grinder, but for native maize too.

Hopi Indians are a peaceful people with a long and often troubled history. For centuries before the arrival of the Europeans they were persecuted by the Navajo, whose vast territory straddled eastern New Mexico, northern Arizona and southern

Utah. Subsequently, the invading Spanish and other European colonialists practised their own ethnic cleansing, made worse by devastating droughts over several centuries, which decimated the Hopi population. Today the sovereign nation of the Hopi tribe is on a 1.5-million-acre (600,000-hectare) reservation in the middle of the Navajo Nation in northern Arizona. They are a very private people and visitors to their villages are not permitted to record them in any way other than with the written word: no photos, no video or sound recording, no painting. The Hopi have been pastoral for centuries, relying on the 'Three Sisters', to provide their subsistence diet. They are arguably the most skilled of all Native American farmers and their ability to develop a sustainable system of agriculture in one of the most challenging and inhospitable parts of the world is nothing short of brilliant.

A Quest for Maize

It was September and I was on a 2,500-mile road trip that took me from Phoenix, Arizona, first to Winslow, a town immortalised by the famous line in the Eagles song, *Take it Easy*. The town is a major railhead with a never-ending succession of enormous freight trains hauling between Chicago and Los Angeles. Amtrak also use the line for passengers, and the Hotel Posada doubles as the town station. So, for an unreconstructed train anorak, this was a place I had to visit. Little did I know, as I checked in and started counting the number of trucks a six-header train was pulling – 142 as it so happens – that my hunt for native Blue Hopi maize was to begin in the hotel restaurant, which was run by Englishman John Sharpe.

On the menu was a traditional flatbread called piki, which is rather like a crispy form of filo pastry made using blue maize. John wasn't in the kitchen that night. Not for the last time would this be a journey of missed encounters. I had hoped he might furnish me with an introduction to his Hopi suppliers, so I could learn more about their amazing crops, but I continued my journey east to Santa Fe next morning, unsure if I would ever meet a Hopi farmer.

Serendipity remained my faithful companion. I was staying with a cousin in Santa Fe who had worked as a GP at clinics in many of the Native American pueblos (villages) and communities in southwest USA. She had a contact who might be able to help me. Melanie Gisler is the Southwest Programme Director for the Institute of Applied Ecology in Santa Fe. This is an organisation dedicated to conserving native species and habitats and the cultivation of native foods. Melanie introduced me to a quite remarkable man, Emigdio Ballon, who is the Agricultural Resource Director for the Pueblo of Tesuque, a community of about 500 indigenous people, a few miles north of Santa Fe.

Emigdio is a busy man. He was about to jump on a plane to Italy to attend a Slow Food conference, but he graciously agreed to show me and my wife Julia around the farm that he was re-establishing at the pueblo. The farmers and gardeners there grew the Three Sisters among many other crops, including chillies, sunflowers and various brassicas and potatoes, as well as a vast suite of herbs that are used in native medicine. One of the most exciting things he showed us was the seed bank he had built. It is home to a large number of local varieties that are being grown on the pueblo as well as other seeds from native communities further afield. There was the Hopi Lima bean, yellow and red pinto beans, the Anasazi bean, which has been grown by Native Americans across the southwest for centuries, and the Golden Tepary bean, cultivated by the Tohono O'odham tribe of southern Arizona.

The climate of this corner of New Mexico is challenging to say the least; a land of frosts from October to May, little rainfall at the best of times, and brutal but short hot summers. So local varieties have become adapted to extreme weather thanks to diligent farmers and careful selection over generations. Many of the crops at Tesuque were of Hopi origin, including both blue and white maize. One of the elders on the pueblo, Michael, talked about Hopi Blue maize as 'good medicine', beneficial for one's health. He made a sloppy porridge with coarsely ground maize for breakfast. It was from Michael that I learned that Hopi seeds of many crops were being grown widely by Native American farmers.

I had hoped I could charm my hosts into giving me some seed from the seed bank, but I was to be disappointed – not because they didn't want to share seeds, but because the new harvest was yet to be cleaned. So, despite my best efforts, I left Santa Fe empty-handed. Thankfully, the Tesuque seed bank is there for all who are keen to grow and save Native American varieties. It is also acting as a low-tech model for other native communities to develop their own seed banks, for it is only by growing and disseminating seeds that these socially and culturally significant crops will survive.

The Quest Continues

It was a long but magnificent drive to the magisterial Canyon de Chelly, deep in the heart of the Navajo Nation in northeastern Arizona. In the bottom of the canyon, with its precipitous red cliffs and ancient cave dwellings, farmers have been growing a wide variety of crops for centuries. Famous for their peaches and other stone fruit cultivated from stock introduced by the Spanish in the seventeenth century, the Navajo of Canyon de Chelly had seen their orchards almost completely destroyed by the US army in 1864, one assumes in an attempt by the invaders to starve out the locals.

Maize was first grown in North America about 4,000 years ago and was a staple of the Navajo then, as it continues to be today. I was too late to see it in the amazing irrigated fields on the canyon floor as the harvest was long past and, in the short time I was there, I did not manage to befriend a farmer either. Understandably, the bitter history of European exploitation and genocide, the many wars fought by the Navajo and the appropriation of their land has left an indelible stain on relations. I had more work to do if I was to be able to talk to any Native American about our shared interest as growers. This road trip, this seed hunt, was not going well. The positive message I got from the journey, however, was that the indigenous population was actively farming and maintaining the cultural heritage of their food crops. Good news.

The following morning, we hit the road early. Although it was only a few hours' drive to Page, our next overnight destination, we wanted to spend some time traversing Navajo country, passing through a couple of Hopi communities. First stop was one of the oldest trading posts in the Southwest, Hubbell, in the town of Ganado, 40-odd miles south of Canyon de Chelly. Now a National Historic Site, it's been in business for well over a hundred years and is a mecca for anyone wanting to buy the finest-quality Navajo rugs and jewellery. When the founder of the trading post, John Lorenzo Hubbell, first set up shop he planted orchards and grew vegetables from commercial varieties available in the US at the time, as well as local varieties being grown by the Navajo. I was able to buy white Navajo sweetcorn which has thrived in Wales and is my absolute all-time favourite, being sweet, flavoursome and at its best lightly steamed. In addition, joy of joys, was a Chinese-made, hand-operated corn-grinder: even today an essential tool in the homestead kitchens of Americans who like to grind their own corn. Now, finally, it might be possible to make those tortillas of my dreams. But first I needed to find some native blue maize.

As I was making this road trip through the American Southwest in the autumn of 2018, I reflected upon some of the wonderful work that is going on across the US for the conservation, preservation and promotion of local heritage and heirloom foods. The native cuisine of the Southwest is the oldest in North America. For over 3,500 years, until the arrival of Europeans in the middle of the sixteenth century, there were about 20 local tribes who had developed a variety of agricultural practices and approaches to crop development. These had originated in Mexico and been honed in the widely variable growing conditions of this part of the USA. I had to be reminded on this trip that across the upland plateaux of Arizona, New Mexico, Utah and Nevada, despite daytime temperatures even in mid-winter in the mid-20s Celsius (mid-70s Fahrenheit) and summertime temperatures consistently above 40°C (104°F), there are frequent frosts for eight months of the year which can have a devastating impact on crops. This makes the climate in my corner of South Wales positively benign by comparison. The growing season in this part of the USA is short and unforgiving. Rainfall is sparse at the best of times and drought has been a persistent destroyer of crops and livelihoods since Native Americans first started to farm here. I was in awe of the horticultural skills of my fellow growers in this beautiful part of the world.

Highway 264 runs east to west through northern Arizona, the southern heart of the Navajo Nation. A landscape of vast rolling plains which, with every rise, reveals another desiccated landscape of canyons, ridges and buttes. Ribbons of trees growing alongside empty washes of dried-up beds await the nourishing fury of the next flash flood, which will arrive during the brief rainy season of high summer. The roads were empty, straight as a die through the basins and then sweetly curvaceous as we climbed the ridges. At each summit we were presented with stunning views of a land that seemed to go on for ever. There was no sign saying 'You are now entering the Hopi Reservation' as we crossed the line between Steamboat and Jeddito, settlements of rickety mobile homes and ancient pick-ups. Hopi tribal land has three mesas at its centre. *Mesa* means 'table' in Spanish. It is the perfect description of these fabulous mountains of the American Southwest: great stretches of flat-topped ranges with precipitous sides. The mesas run north to south, each separated by miles of flat basin land.

A Little Luck

Blink and you miss First Mesa. Behind the modest houses are a private people who prefer it, I think, if travellers just pass by. In contrast, the Second Mesa had a school, bedecked with flags and bunting as that day it was open for visitors to enjoy seeing the children performing their traditional dances. A warm and welcoming place.

As we ascended the mesa and drove onto the top, we passed a string of small shops that were mostly selling Hopi jewellery and trinkets. My destination was the Cultural Centre. I fancied a blue Burrito for lunch and hoped to find some seeds there, as was the case at Hubbell. The Hopi lady at the checkout told me that my best chance was to go to a store a half-mile back along the road which sold blue maize seed. But could I find the place? We drove up and down a couple of times until finally, exasperated, I pulled into the yard of a gift shop, called *Tsakurshovi*, meaning 'The hill that comes to a point' in Hopi. Behind the wooden counter an American was chatting to a Hopi working alongside him. Joseph Day was looking at an article about the domestication of maize. I asked him if he could point me to the store that sold Hopi seed and he laughed, 'You've come to the right place for Hopi seeds! Not that store which sells maize grown using irrigation!'

Joseph's shop is the unofficial cultural centre. He is originally from Kansas. He had met and married a local Hopi, Janice, some 20 years ago. Together they had returned to her village above the Second Mesa. Hopi is a matrilineal society, in which property and inheritance pass through the female line. Janice's grandmother had cultivated a swathe of ground behind their shop which had been left to nature. When the couple returned, Janice asked the elders of the village if she could take back her family's land to farm. The answer was: 'Sure. Just remember: no fences and be sure to grow Hopi crops.'

Janice came into the store, her hands dusty with flour. She was in the middle of baking, but was happy to give me some of her Blue Hopi seed and some white Hopi Tepary beans too. I was overjoyed and so grateful. I told them both how I had grown Hopi maize and a speckled Tepary bean in my own garden. My maize did well but was very tall. The maize Joseph grew was no more than four feet high when mature, but, then again, his garden of corn, deep in the arid American Southwest, could not be more different to mine, nestled in one of the wettest parts of the U.K.!

The traditional Hopi way to cultivate maize without irrigation is completely different to the close-planted serried ranks we see growing across swathes of our lands. The Hopi grow maize in clumps about 2.5m (8ft) apart. A few seeds are planted, as much as a foot (30cm) down, to make the most of what little moisture there is. Sometimes the seedlings are thinned to three or four plants. The habit of the Hopi's maize is perfectly adapted to arid conditions. The leaves of the plants are long and broad, so they shade the ground they are growing in, thus helping to retain moisture. There is considerable genetic diversity in traditional maize varieties, which makes it possible, in the hands of great farmers like the Hopi, to select the best traits to suit different climatic conditions. Hopi Blue maize is no exception. The cobs are not uniform and are able to grow with no irrigation other than what rain falls from the sky during the brief and unreliable rainy season in July and August.

I get a bit emotional when I am given heirloom seeds from sources which are so precious and rare. There are several varieties of blue maize that range in colour from steel-blue to almost black. With blue tortilla chips and flour being very popular, the commercial development of blue maize has been very successful. However, this has presented a threat to the uniqueness of local or Folk Varieties (FV) such as the ones Janice gave me.[1] Hopi farmers have found it difficult in such challenging growing conditions to maintain a consistent quantity of their favourite types. So, many have bought commercial varieties that are botanically synonymous with (similar to) the FV ones in the hope they will yield better crops. When the bought-in seed has proven to be very close to the original, farmers have focused on growing that in preference to the FV seed. For this reason, being able to save all the myriad landraces and variable

206

traits is an uphill struggle. However, like all farmers, the Hopi want to cultivate crops that are best suited to their land, so the process of continual selection by saving their own seed, including from commercial varieties, is a way of life.

Hopi farmers are known for being very enthusiastic experimenters, growing and comparing seed from a variety of sources, including modern cultivars and FV from other indigenous farmers.[2] Traits that make an FV more stable – for example, germinating even when sown as much as a foot deep – are a specific adaptation to the very dry conditions and poor soils that Hopi farmers must deal with. The Hopi people's relationship with corn is fundamental to their identity. It is sacred. It is their staff of life. The colour of their crops is of great importance too, and only seed from ears that show a pure colour characteristic are saved to be sown the next season. Responsibility for the preservation of distinct varieties is handed down from one generation to the next.[3] Today, there are still as many as 17 different varieties of maize being grown in the Hopi Reservation. Their constant curiosity and continual experimenting with selection and cultivation, a tradition that goes back thousands of years, resonates with how I imagine the earliest farmers worked on improving their crops.

In her book *WorkaDay Life of the Pueblos*, published by the US Department of the Interior in 1954, Ruth Murray Underhill, when describing how the Native American and indigenous peoples of Central and North America maintain varietal consistency, observed how Hopi farmers were most proprietorial with their seeds, with no one farmer growing all the 19 different varieties she observed under cultivation. The only time the seeds were shared was when a son got married. He would bring with him to the new marital home some of his finest seed to plant on his wife's land. Underhill suggested that being the son of a farmer known for the quality of his maize would make him a very desirable catch!

Knowing this, I better understood how very lucky I was to have been given seeds by Janice that hot September afternoon. I remember too, when admiring the seed bank at the Tesuque pueblo in Santa Fe, that Native Americans, including the Navajo, would go to Hopi farmers for seed because they produced crops of the finest quality and were the best suited to local growing conditions.

A Truly Man-Made Vegetable

The story of the domestication of maize and the unravelling of its genome is one of the most remarkable stories of the genius of the very first Neolithic farmers. Botanists need to know who the wild parents – the Adams and Eves of the plant kingdom – are when investigating the genome of domesticated crops. Rich in genetic diversity, decoding their genetic make-up is crucial to understanding the process of domestication and in developing methods for breeding new varieties.

'Maize domestication is one of the greatest feats of artificial selection and evolution, wherein a weedy plant in Central Mexico was converted through human-mediated selection into the most productive crop in the world. In fact, the changes were so astounding that it took much of the last century to identify modern maize's true ancestor.'[4] So wrote the geneticist Feng Tian in his introduction to a research paper on the domestication of maize while working at Cornell University in 2009. Until the 1930s scientists had been unable to identify the wild parent of maize. It was one of the great mysteries of domestication.

The family of grasses called Poaceae, of which maize is a member, includes wheat, rice, barley, oats and sugar cane.

The genus *Zea* is specific to a group of grasses from Central America and Mexico. Domesticated maize's full botanical name is *Zea mays* subsp. *mays*. Just how had the first farmers of Mexico 10,000 years ago or more been able to breed such a hugely diverse number of varieties of maize and, as we shall see, in such a short time? Who were the wild ancestors of one of the world's most important food crops?

Until Darwin revolutionised biological science, there had been a long-held misconception that evolution is always slow and gradual. He was minded to challenge this, but it was to take many decades for the opinions of others to change. The fact is that minor changes to influential genes can produce dramatic and speedy evolutionary outcomes. This was never better illustrated and argued over among scientists than with maize which, as if by magic, suddenly appeared in the archaeological record. The prevailing view of evolutionary change being slow and gradual, which continued well into the twentieth century, meant that plant scientists were struggling to find an explanation for the magical appearance of maize. Where to start looking?

The Search for Its Wild Parents

There is another group of wild grasses included in the same genus as maize, called teosinte, whose full botanical name is *Z. mays* subsp. *parviglumis*. The problem scientists were presented with is that teosinte and maize may be of the same genus, but there the similarity ends. Teosinte is a short weedy grass that, unlike maize, doesn't have a cob-like 'ear'. The head consists of a maximum of a dozen kernels in two rows that leapfrog along a hard outer stem which protects them. Maize, on the other hand, can have up to 20 rows of exposed kernels. Nineteenth-century botanists considered teosinte and maize to be so unrelated that they gave teosinte its own genus, *Euchlaena* (now assigned to a species of moth). Geneticists have spent the last hundred years using the latest advances in genetics to understand in what way these two utterly different-looking plants are, in fact, intimately related.

The huge difference in the morphology of teosinte and cultivated maize led two American researchers, Paul Mangelsdorf and Robert Reeves, to publish a paper in 1938 hypothesising that maize was domesticated from a now-extinct wild maize and that teosinte was a cross between a common crop called gamagrass* and this imagined extinct maize.[5] This hypothesis seemed to be confirmed when Mangelsdorf and Reeves were able to successfully cross gamagrass with maize, although the resulting offspring were sterile and only came into this world thanks to some fancy botanical lab work. Job done supposedly except, in the world of scientific questioning and scepticism, academic fisticuffs continued. It took a Nobel laureate to come up with a better solution a year later. George Beadle was not persuaded that teosinte and maize were so different that maize needed an extinct relative to be with us today. In his seminal paper published in 1939, *Teosinte and the Origin of Maize*, he hypothesised that it was the ingenuity of neolithic farmers 10,000 or more years ago identifying and selecting traits in the genetically diverse teosinte that enabled rapid mutations, each

---

* *Tripsacum dactyloides*, commonly known as gamagrass and widely grown in the Americas as a fodder crop, is the cultivated sibling of a wild member of a tribe of grasses Maydeae, which includes cultivated maize. A tribe in the classification of plants is defined as a further division between a family or sub-family of plants which share a number of specific characteristics.

causing relatively large changes.[6] It wouldn't have taken long – just a few generations – to end up with a maize remarkably similar to the modern varieties we all depend on today. Indeed, Beadle turned the work of Mangelsdorf and Reeves on its head, suggesting that just four genes, each controlling a single trait, differentiated teosinte from maize. He also dismissed their breeding attempts, which were successful only because of great intervention and resulted in plants that were incapable of surviving in the real world.

Since the end of World War II, a huge amount of genetic and taxonomic detective work has been undertaken, such that scientists now believe that all maize is derived from a single domestication event with teosinte in the Balsas River Valley, a region of southwest central Mexico. For me, at any rate, that is quite an amazing thought – some 10,000 years ago Neolithic farmers were harvesting wild teosinte and selecting seed from varieties that had more kernels, grew more abundantly, and probably tasted pretty good too. Just how long this actually took is now being determined through the latest genetic sequencing techniques, working with the 20,000 landraces of teosinte and maize that are held within gene banks and research centres around the world. What is clear is that it didn't take many generations of selection before some farmers hit the jackpot. The result of their persistence led to a process of selection over generations of farming across Mesoamerica and parts of South America, arriving at the countless varieties we enjoy so much today. Some story.

More than One Maize

There are six distinct types of cultivated maize, the oldest of which is Indian or flint corn, *Z. mays* var. *indurate*. It is distinguished by its hard outer seed case and kaleidoscopic range of colours. Its most common use was and continues to be to make hominy, the name given to the whole kernels after nixtamalization – a process of soaking in lye (sodium or potassium hydroxide). For thousands of years this was produced by mixing water and wood ash together and had the effect of making the kernels more nutritious and digestible. Because of its many colours this corn is also an important part of Native American tribal rituals and belief systems.

Dent corn, commonly known as field corn, *Z. mays* var. *indenta*, is grown primarily as a fodder crop or for food processing, including oil production and as a biofuel. It comes in two colours, white or yellow, and has kernels that contain both hard and soft starches, which causes the seed to become indented when mature and dry. *Z. mays* var. *saccharata* and *Z. mays* var. *rugosa* are both forms of sweetcorn. Another type of flint corn with a super-hard outer shell and very soft, starchy inside is popcorn, *Z. mays* var. *everata*. When the kernels are heated up, the moisture inside turns to steam, causing them to explode. The result is a fluffy blob of starch blown inside out. Archaeologists have found evidence of this type of flint corn in New Mexico from 3,500 years ago. I imagine the earliest native settlers, weary from the long journey north from southern Mexico, relaxing by the campfire enjoying a bowl of freshly popped corn drenched in honey harvested from a swarm of wild bees. Perhaps this is the USA's oldest culinary tradition? Sweet or salty, it is certainly one of its most pleasurable. Finally, there is flour corn, *Z. mays* var. *amylacea*. It has soft, starch-filled kernels which are easy to grind and has been a staple variety for Native Americans for millennia. Although mostly white, flour corn comes in several different colours, including the one that means the most to me, blue.

The Rise of the New

My journey through the Hopi Reservation reminded me of an encounter in Shan State, northern Myanmar, which had highlighted a different existential threat to Folk Varieties. I had visited one community of farmers living near Hsipaw, a small town on the road from China to Mandalay, on three occasions and found the women stripping the sheaths off great piles of modern hybrid maize cultivars. They had grown the crop from seed supplied to them by Chinese middlemen who were giving them a guaranteed price for a crop destined for Chinese consumers.

Shan farmers have been growing maize since it was introduced to Indo-China by the Portuguese early in the sixteenth century. As a result, the country used to be rich with a diverse range of colourful, nutritious and well-adapted local varieties, but now farmers were almost exclusively growing these Modern Varieties (MV). The only place to find examples of the dwindling FV crop was in the market where, before dawn, women were steaming a local variety for hungry punters to eat as a breakfast on the go. Although not as sweet as modern hybrid corn, the cobs were nutty and delicious. I fear that, like other native crops that have appeared in these pages, Shan maize will soon be lost forever unless local varieties are either grown to enrich local cuisine or held in a seed bank. I feel confident that Hopi maize is safe for future generations because Hopi farmers are passionate about growing it and more and more Americans are coming to value it as a cornerstone of their indigenous food culture.

Over the centuries of adaption and selection that started from the moment the Portuguese planted crops in their colonial outposts, varieties of maize have been bred that are each uniquely adapted to cope with a very wide range of climatic conditions. Today, modern cultivars and new varieties of GM maize have enabled it to be grown in climates that were once considered most unsuitable – my own British climate being one of them. As is evident in South-East Asia, including Myanmar, these modern cultivars are replacing local ones on continents that have been growing their own varieties for centuries. We see similar things happening in many parts of Africa that are suffering particularly from climate change and drought. One of the great advantages that FV crops have over MV types is that they are more genetically diverse.[7] This diversity can be present in single plants. Known as heterozygosity, it is key to the farmer's process of selection. Greater diversity is synonymous with crop stability, meaning less variation in yield from year to year. Both theoretically and in practice, low-input indigenous farmers growing local varieties experience a much-reduced risk of crop failure due to unexpected climatic events like flood and drought. This reduces the need for unaffordable high inputs, including chemical fertilisers, herbicides and pesticides. Most importantly, farmers have access to a reliable source of well-adapted saved seed for little or no cost which they can sow in the next season.

Modern agricultural practice usually involves working with crops that lack the genetic diversity of indigenous varieties. This relative lack of diversity often means that, in the event of a shortage of costly chemical inputs needed to grow them, industrial farming is more vulnerable to environmental conditions and therefore experiences more yield variability.[8] For me at least, this begs the question: why are we encouraging indigenous farmers, who are for the most part living a hand-to-mouth existence, to abandon their traditional and dependable agricultural practices in favour of modern varieties of maize that are most prone to climate extremes?

The dangers of catastrophic crop failures due to relying on an increasingly narrow genome in maize has already been experienced in the USA. There is an ongoing debate about the sustainability of GM maize cultivation when industrial-scale production uses up vast resources of water and fertiliser at great expense to the taxpayer and the environment.[9] The drought of 2012 had a devastating effect on maize production, which fell by more than 35 per cent. GM maize is also in the dock for failing to live up to the hype of improved yields and resistance to pests and diseases. The protagonists endlessly argue over statistics, but the reality is that unless the world diversifies its food production, a single climatic event could prove catastrophic for us all.[10]

Whenever I walk among the ripening ears of corn in my garden and select a couple for immediate consumption, I am repeating an action that has been performed countless times since the dawn of settled agriculture in the New World. To me, maize embodies all that is exciting and rewarding about plant breeding: from the folk variety of Blue Hopi maize that is at the very centre of Native American culture now growing in my polytunnel to the maize flour I bake with. And not forgetting the delicious modern hybrid cultivars of 'super sweet' sweetcorn developed to thrive in our cooler and wetter climate, and the maize oil I cook with. Maize in all its forms is now a vegetable the world really cannot live without. Today, there is even a new variety of open pollinated miniature sweetcorn called Blue Jade. But nothing compares with Janice's precious gift to me, her family heirloom Blue Hopi maize.

The Tale of
Two Classy Beans

...they must be young and freshly gathered:
boil them till tender, drain them, add a
little butter and serve them up

Mary Randolph (1762–1828) –
The Virginia House-Wife (1824)

T he winter morning in Hsipaw, a small town on
the road to Mandalay from China, was chill. It
was February 2015, and I was travelling in Shan
State, northern Myanmar. Mist lay across the silently
flowing Myitnge River like a comfy duvet. The lightening
sky drove out the shadows that had been hiding the secrets
within the morning market. Stooping beneath the gently
billowing, red and orange fabric canopies of the stalls, I
was soon lost in a maze of narrow paths, muted colours,
spicy aromas and piles of local produce. Everywhere were
mounds of watermelon, pineapple and the usual spread
of chillies, onions, greens, herbs and salad crops. I wasn't
looking for anything in particular – just curious, with the
hope of being surprised.

A Very Angry Bean

On the face of it there was nothing strange about seeing a diminutive lady of indeterminate age squatting on the road-side at the edge of the market with a mound of freshly cooked beans for sale. Having short pods with three beans in each, they looked to me like lima beans, *Phaseolus lunatus*, which are native to Peru. People were buying cupfuls and thumbing the young beans out of their pods, as one does with edamame beans. In Hsipaw they were an early morning snack on the go. My detective nose began to twitch. This bean required closer inspection. Those being sold that morning were pale green, with their skins speckled red. I bought a portion of the freshly cooked beans to try. They had a distinctive chestnut flavour and meaty texture. Rather good, I thought. Time to consult my guide, Ah Soe, a native of Hsipaw. He told me that the bean was local to Shan State and was called *Tow Lai Se* – the 'Angry Bean'. It was so named because tradition had it that if a pregnant woman attempted to harvest the bean, she would anger it and it would no longer grow. Such a strong cultural identity with a bean that I presumed came originally from South America I found profoundly intriguing. The Angry Bean was clearly deeply embedded in Shan culture. The Shan people are Buddhists and bring with their beliefs a powerful animist tradition. The negative image of pregnancy was redolent of the belief that menstruating woman are 'unclean'; a belief still held by some monotheists. I imagined that when the bean first arrived in this part of northern Myanmar, the result of globalisation with the opening up of new sea trade routes early in the sixteenth century, it would have been seamlessly assimilated into the local cuisine which already included the similar-sized fava bean. It would not have been seen as a new arrival, perhaps merely something familiar that had changed a little – just as we have seen with the common bean's arrival in Europe, which was assumed to be local until geographers and botanists recognised clear differences in morphology. The striking appearance of the Angry Bean might well have been the catalyst which started its journey into Shan tradition.

I was desperate to acquire some seeds, but I had a problem. The beans being sold were all young and fresh. There were no mature ones to be found. Fortunately, Ah Soe came to my rescue. His mother had been a keen gardener and she might have a few seeds left over from the previous year. The next day he showed up at my hotel with just five beans. They were stunning, each the size of my thumb and a wonderful mottled cream and red. I had never seen anything like them before. His family had been growing them for generations, but sadly no more: they were upsizing and their garden was now a building site.

I returned home with my precious cargo and was able to grow a modest crop in a large pot in my greenhouse, which ensured that I had enough seed for the following year. It's a bean that Jack would have liked. My plants grew to over three metres and later crops have taken over my greenhouse.

The Angry Bean was clearly considered native by Shan people and an intrinsic part of tribal cuisine and culture. The name alone suggested a long history and powerful link to traditional beliefs. The place of the lima bean in Burmese agriculture has a long tradition. Myanmar is a major exporter of all sorts of beans to India and China, including a type of small, white lima bean known as the sieva bean. I had seen plenty of these beans being sold across the country, unlike its close relative *P. lunatus* subsp. *pallar* which I only ever saw sold in Shan State. Tony Winch in his book, *Growing Food*,[1] describes the sieva bean as also going by the name of Burma bean and the Rangoon bean; so named because they were grown extensively in that part of the world. Clearly, both types of lima bean were valuable but for different reasons.

How had the bean I had found in the market in Hsipaw become an intrinsic part of Shan culture and was it a type of lima bean as I suspected? I tweeted a photo of dried Angry Beans, asking if anyone could shed light on what they might be. It was not long before a fellow seed saver, Annie Wafer, who lives in Slovenia, told me that an elderly lady in her village grew them and sold a few in the market. Apparently, she had brought them from America when visiting her family there. Annie suggested my bean *P. lunatus* subsp. *pallar* could be the Christmas Lima bean. And she was right. It is included in Slow Food USA's Ark of Taste and is part of US food history.[2] So, this glorious looking and tasting bean is embedded within two completely different cultures on opposite sides of the planet, Myanmar and the USA. But how?

Deadly Parents, Delicious Offspring

Lima beans were widely cultivated in South America at the time Columbus first arrived in the New World. Jonathan D. Sauer[3] describes their wild relatives as being small and highly toxic, as they contain up to twenty times the concentration of glucoside that the cultivated varieties do. This chemical breaks down into hydrocyanic acid – cyanide to you and me – when bruised or chewed and is therefore poisonous and possibly fatal, unless made safe by prolonged boiling in many changes of water. This begs the question: what motivated the indigenous people of South America and Mesoamerica to eat such a poisonous plant? It must have taken many deaths and much experimentation for those Neolithic farmers to figure out how to make the bean edible and then, through careful selection, develop varieties that contained less glucoside. Proof, if more was needed, of the genius of our distant ancestors. Even today lima beans should never be eaten raw because they still contain small quantities of glucoside, something shared by most beans, which can cause serious stomach ache.

The wild lima bean is common in Central America, especially Guatemala on the Pacific Ocean side with its wet and dry seasons. Its range extends north into Mexico, and it can also be found in remote corners of Belize, Venezuela and Puerto Rico. Colonies have also been reported on the eastern slopes of the Andes of Peru, Argentina and Brazil, but there are no historical records of the domesticated lima bean within the reported range of its wild parent. This suggests that the first domesticated beans moved across the continent with the communities who grew them, whereas the local population probably continued to forage the wild beans. In recent years, plant breeders have been able to cross-pollinate wild with cultivated varieties, to develop new cultivars: in a sense their work replicates in part what those migrating communities were doing themselves.

The archaeological record indicates that the lima bean was domesticated in the northern regions of its native homelands: Mexico and Puerto Rico and also some parts of the western regions, principally northern Chile and Peru. However, the varieties from the two regions are quite different. Those from the north are smaller, like the white sieva bean that is grown extensively in Myanmar. The first archaeological record of the parent of the larger, locally named *pallar* type, which I found in the market in Myanmar, is in Guitarrero Cave in Peru, dating from around 8,500 years ago. Sauer posits that because the location of the cave is in a deep valley the species could not have been naturally present.[4] Long before the earliest evidence of irrigated agriculture, the native wild parents were probably introduced by settlers from east of the Cordilleran crest, a part of the central Andes. There is evidence to suggest that its domestication happened some 3,500 years later when this large, white and red-speckled bean, *P. lunatus* var. *pallar*, entered the archaeological record as a product of planned agriculture at a place called Huaca Prieta on the north coast of Peru. It was grown in considerable quantities in the irrigated fertile fields that fed the prehistoric Peruvian civilisations. The beans of this period unearthed by archaeologists frequently show their distinct colourings, preserved due to the dryness of the region. The pallar bean also appears as a motif on ancient Peruvian pottery and many archaeologists believe that it had an important spiritual and cultural place in pre-Inca society, most probably because it was also rather good to eat. But there is no doubt in my mind that its metaphysical qualities were identified and celebrated by the Shan people when it reached them 500 years ago.

Sea-Faring Beans

An Old World bean that Columbus would have been very familiar with, and that was carried as ship's stores, was the broad or fava bean, *Vicia faba*. In one of his earliest journal entries, written when on dry Caribbean land, was a reference to a bean eaten by the Arawak that he presumed to be a

221

type of fava bean despite it looking quite different! A cursory glance and one might think the lima bean is like the fava bean (rather as I imagine Shan people did when they first came across it), so I am willing to indulge Columbus's botanical shortcomings on this one. It was, in fact, the sieva type of lima bean. Over the coming centuries this white bean was grown widely throughout southern Europe as well as in the Portuguese colonies of South-East Asia. So it would appear that the sieva bean probably found its way into Myanmar via India early in the sixteenth century. But what about the Angry Bean?

Early explorers and colonisers of the New World used legumes as ships' stores, which is why all types of beans from both the Old World and the New World were soon growing in each other's native lands. Columbus fed peas as well as fava beans to his sailors. For his return journey he reprovisioned with New World beans. As both Portugal and Spain colonised the New World throughout the sixteenth century, they developed regular sea routes to their newly claimed territories. The consequence was that within just a decade or two of the arrival of Columbus, New World beans were being grown widely throughout West Africa and the Indian sub-continent.

It took nearly a hundred years after Columbus first saw the lima bean for it to be catalogued in Europe. Although it had probably been around for decades, it was the Dutch botanist Matthias de L'Obel (1538–1616) who first described both the sieva and the pallar in botanical drawings in 1591.

There were two routes by which crops from the New World could travel to East Asia. One was taken by the Portuguese who, late in the fifteenth century, had already opened up a route to their trading posts in southern India by sailing around the Cape of Good Hope. If they were carrying any lima beans, they too would probably have been of the sieva type because they were grown widely in eastern parts of South America and Mesoamerica. Because the *pallar* types, of which the Angry Bean is one, were grown in Peru and Chile it probably arrived in Asia by a second route across the Pacific: on the Manila galleons. These were Spanish ships that sailed from 1565 to 1815 between Manila, which was the main port of the Spanish colony of the Philippines, and Acapulco on Mexico's Pacific coast. The year-long round trip shipped Mexican silver and Catholic missionaries west into the Asian market and perfume, porcelain, gems and Indian cotton east to the Spanish colonies in the New World and onward to Europe. By the end of the eighteenth century both types of lima bean were a common food crop in China and India. Early in the nineteenth century the *pallar* type had been introduced to South Africa and the colonised islands of the western Indian Ocean. It is reasonable to assume that many of the peoples of Myanmar were growing and eating both types of lima bean in the middle of the sixteenth century, and the Shan people took a particular liking to the *pallar* type, perhaps because it was especially happy growing in the foothills of the Himalayas, which are reminiscent of its Andean home.

Horticulturally, the two types have different growing preferences: the smaller sieva shows great drought and heat resistance, which is why it became a staple for several Native American tribes of Southwest USA. The Angry or Christmas lima bean, on the other hand, does best in nice, damp, mild climates, which is why the Shan people of northern Myanmar liked them. In the latter half of the nineteenth century both types became a major US export and California was by far the world's largest centre of production. Fortunes were made by lima bean farmers in Los Angeles and the surrounding areas. Sieva beans are still farmed in the Sacramento and San Joaquin valleys, but are in decline as farmers replant with vines and other higher value crops.

Commercial lima beans are white whereas the colourful Christmas lima has little commercial value and only survives as a heritage variety in the USA. It can be bought in farmers markets across the western US states. Its future in Myanmar is precarious. When I first discovered this bean there was a number of vendors selling it in the Hsipaw market. However, on my last visit in 2017, I was only able to find one woman with a great steaming cauldron of them. As habits and tastes change, the Angry Bean could go out of cultivation completely and, with it, its unique cultural relationship with Shan cuisine and beliefs. So, I am glad to be growing it and will return seed on my next visit to a new generation of farmers who see the value in reviving and maintaining Shan food culture. Its flavour is unimpeachable. Long may enthusiasts keep on growing it all over the world.

Much as my Angry Bean enjoyed growing in the warm and humid world of Shan State, it only flourishes in a polytunnel in my garden. Like its sister sieva bean, the pallar is no lover of the British climate, which is why it has never found a place in our own food culture. It's a very different story for her close relative, the runner bean, *Phaseolus coccineus*, which is also a native of South America and very similar in appearance.*

---

* The easiest way to spot the difference is in the immature plant. The lima bean does not have a large pair of juvenile leaves when it first germinates, unlike the runner bean. The pods of runner beans are much longer with more seeds too.

A Cool Bean

Unlike lima beans that do well in hot and humid environments, the runner bean prefers things a little cooler, which is probably why it grows so well in Britain. It is native to Mexico and Central America and is found growing at elevations of 1,500–2,400m (5,000–8,000ft). Its wild parents had been foraged by indigenous tribes in the region 9,000 years ago and there is some evidence that it was first domesticated by 4,000 BCE. It was being widely grown in Mexico a few centuries before the arrival of Hernán Cortés in the first quarter of the sixteenth century.[5] Of all the different types of beans that came to the attention of European explorers when they colonised the New World, the runner bean was the last to be brought back to Europe. For them, its pretty red flowers were the attraction because it was first grown as an ornamental named The Scarlet Runner. The original, with its emphasis on flower production, would likely have had shorter, less tasty pods. There is some debate still about the date of its first introduction into England. Might it have been the Jesuit priest John Gerard (1564–1637), who practised his Catholicism covertly in the reign of Elizabeth I, and is said to have grown them up poles in his garden? Elizabeth I is also credited with being the inspiration for the first named variety, Painted Lady – its beautiful, bi-coloured (red and white) flowers are redolent of the monarch's famous chalk and rouge make-up. Or was it the gardener John Tradescant the Younger (1608–1662), who brought seed back from one of his many plant-hunting trips to the American colonies as an exotic addition for King Charles I's garden at Greenwich? Possibly, as there is evidence to suggest that this bean was also being grown as an ornamental in North

America at the time. Today, Scarlett Runner is the generic name used in the USA, where it remains primarily a plant for the herbaceous border.

Flower Power

What is indisputable is that growing runner beans for their tender pods is a British gardening obsession. Only in the last half-century have the Americans occasionally moved the plant from the flower bed to the kitchen garden. Continental Europeans are rather dismissive of the plant as a green vegetable, although its seeds form the basis of much regional cuisine. The French remain contemptuous of the green bean, and you won't find it growing in any self-respecting kitchen garden in France for that purpose. As a shelling bean, yes. It is most commonly grown in other parts of Europe as a dry or butter bean, like its relative the sieva bean.

Yet, to me, the quintessential image of a British kitchen garden or vegetable plot in high summer is the sight of a long line of bamboo canes covered in a tangle of vines, bedecked with bright red or white flowers and slender, brilliant green beans. Come August, gardeners are offloading armfuls of their harvest to all takers as we seem incapable of growing just a few. A glut is inevitable and there comes a point when most of us who love eating runner beans simply cannot face another mouthful.

I particularly enjoy growing two delicious continental heritage varieties for their beans rather than their pods. The Spanish call runner beans grown for their seeds 'Judion', and my favourite is the enormous Judías Di Barco de Ávilar, which has protected status. It is as delicious as it looks, the perfect ingredient for a great soup or stew. The French call these types of bean *haricot blanc*. My favourite French variety is Gros de Soissons. There is also a variety of climbing French bean that is unrelated but looks identical and has the same name! The Italians and Greeks have their own large, white butter beans too, each part of a distinctive regional cuisine. Fasiola Gigantes has protected status in northern Greece and forms the basis of many traditional Greek recipes. Today there are many modern cultivars with white flowers too, including the Czar, a British classic famed for both its long and tender pods and its large, white, nutty seeds.

A Long British Tradition

Every year a strange ritual takes place in village halls across the land. The Flower and Produce Show has been a major annual event for rural communities for at least the last 200 years. Picture the scene.

The late summer sun streamed through the tall windows of the village hall and lit up tables laden with produce like divinely illuminated altars to fruit and vegetable perfection. The tension was palpable. This was a competition with just a thin veneer of civility. Despite those present oozing neighbourliness and polite curiosity, this was an event where, faux protestations aside, amassing rosettes and cups really mattered. The usual suspects were present. A convivial bunch of newbies showing the fruits of their gardening labours for the first time, rubbing shoulders with hardened old-timers who would take the secrets of their success with them to the grave. Like me, everyone wanted to win! Despite having had a vegetable plot for the best part of six decades, this was only my second time of participating in this seasonal event. The idea that anyone could grow a vegetable more perfectly than me is something I am too vain to be reminded of. However, for once, I had weakened under pressure from fellow villagers to enter into the spirit of things and compete. So, laden with goodies, I started to set out the various vegetables I had hoped were good enough to win, including my longest runner bean.

Well, it wasn't to be – not that year at least. Yet again the prize would go with ease to Norman. He had remained unbeaten since returning to the village some 20 years ago. So, what was the secret of his success? A bean named Stenner.

The Stenner Story

Runner beans can cross very easily with other varieties growing nearby because the flowers are pollinated by bees. So, it is necessary to isolate the crop if one wants to save seed which is true to its parent – not easy if you have an allotment or neighbours who share your passion for this fine vegetable. In my case, when I grow runner beans for the Heritage Seed Library, I try to persuade those living nearby to grow the same ones. Sometimes this is easy as my fellow runner-bean-lovers are a competitive lot and never stop dreaming of lifting the trophy for the longest runner bean, thus denying Norman yet another year of triumph. And of all the many varieties I have in my library there is one true prince: Stenner, Norman's bean, is in a class of its own. It was bred by Welshman Brython Stenner who lived in South Wales. In the 1970s he noticed that one of his plants of a very well-known variety, Enorma, had particularly long and fine beans. Over some years of careful selection, he bred a bean that was not only very long, up to 40cm (16in), but delicious too. The Stenner bean reigned supreme as national champion for the best part of 20 years and was widely grown by enthusiasts as a show and competition bean. Sadly, Mr Stenner died in 2002 and some of his surviving seeds were passed to the HSL by his family. Bean growers, like most plant breeders, are never satisfied, so even though Stenner can still be found on the show bench, new varieties have been created, including the appropriately named Jescot Long-Un, the longest bean I have ever grown, bred by a competitor in the South East of England in the 1980s.

I grew Stenner for the first time in 2007 and, the only other time I have competed, won the prize for the longest bean at my local show. In 2018, the HSL asked me to grow out seed for them again and sent me sufficient from the 2015 crop that I had returned to the library, such that all my neighbours living within a kilometre could grow it too. There was much excitement that at last there would be a level playing field. We would all compete for the longest bean on equal terms, and it would be down to our horticultural skills as to who would lift the coveted longest runner bean trophy. Despite my best-laid plans there was trouble ahead.

Norman was a professional plant breeder and acquired Stenner seed from the great man himself over 20 years ago. I was keen to see if his beans had changed over the years because his garden adjoins others where different varieties have been grown. The bad news was that, after comparing his bean pods and mine through the growing season, it was obvious they were different. The good news was that Norman's own beans, when closely examined, were identical to the description of the variety held by the HSL. So, what had gone wrong? My beans tasted as good and were fairly long, but the difference in morphology showed that I had clearly failed to maintain genetic purity: the result of accidental cross-pollination with a different variety growing within half a mile or so of my garden. At least I could eat them but none of the seed was fit to be saved for the library. Whether by luck or judgement Norman's beans had not suffered the same fate as mine. Thanks to his diligent method of selection, I would subsequently grow some of them to restore purity to the library. That summer he would walk away with the trophy yet again and I was not brave enough to tell all my neighbours they had been growing a rogue bean.

A Sense of Place

A word more often associated with wine than vegetables is *terroir*. It is used to describe all the environmental factors – its growing environment, location and method of cultivation – that affect a crop's appearance, colour, habit and other obvious traits, including, most importantly, taste. Together these various influences give a particular variety its own character. *Terroir* is used as the basis for giving all sorts of foods and drinks special status in order to protect their production and unique place in a nation or region's culture. It started in France with wine and the creation of *Appellation d'Origine Contrôlée* (AOC), a system for regulating and protecting wines both in France and around the world. *Terroir* has now been applied to a number of crops, including beans, chillies, tomatoes, certain heritage wheats, chocolate and even tobacco.

So, in what way is this relevant to the status of the runner bean? Because of its promiscuous nature it is very easy, either by design or by accidental cross-pollination, to create new varieties. Runner beans only became popular in the kitchen garden in the nineteenth century and then there were very few varieties in cultivation. As the story of the Stenner bean shows us, accidental crossing or mutation can deliver something different. However, all runner beans are in reality very closely related and all those 'home-saved' beans that have been grown again and again over the generations across the country have their own *terroir*. This is very evident in my corner of the U.K., South Wales.

It's All in the Name

Most runner bean seeds are purple and speckled black in varying degrees. Some varieties have all white seeds, while some are all black. These variations are due to tiny changes in an allele – a variant of a given gene – which affect the colour of the seed, much in the same way as alleles affect the colour of our eyes. South Wales seems to be particularly blessed with black-seeded heirlooms. I first became aware of them when my friend Liam Gaffney sent me a runner bean from South Wales called Brecon Black. It is held in the Irish Seed Savers Exchange along with another South Wales heirloom – although with speckled seeds – named Yardstick. A very nice bean it is too. Subsequently, after giving a talk to a group of local gardeners, a member of the audience gave me a packet of seed bred more than half a century ago by a keen competition grower, Welshman Allan Picton. He named his bean Rhondda Black after the valley in which he lived. It is the result of an accidental mutation of the same bean, Enorma, that begat Stenner. After 20 years he wanted to 'improve' Rhondda Black with an even longer and straighter specimen, so he crossed it with Jescott Long-Un. It certainly gives Stenner a run for its money, both in length and flavour.

Another black-seeded bean that was once grown across the region, known as The Miners' Bean, sadly now appears to be extinct, although I would not be surprised to learn that it still flourishes in some gardens and allotments in the Welsh valleys. One need look no further than my corner of South Wales to see how deeply embedded the runner bean is in British gardening life. The HSL holds another variety, Pwlmerick, named after the hamlet in which it is grown, just a stone's throw from my own garden.

My library is full of runner beans that have a particular *terroir*. The prolific and delicious Montacute bean, grown for generations in the garden of the stately home in Somerset after which it was named, is just one. Naming our crops makes them part of us. I encourage gardening groups and communities to name beans they grow from their own saved seed. Over the years, regardless of where the first beans came from, they will have shared their genes through multiple cross-pollinations. These beans become landraces – locally adapted versions of a variety, with inherited genetic traits that predispose them to grow in their 'neck of the woods'. Whether they are tastier, longer-podded, shorter, grow taller or flower early is of less importance than their connection to the local community. Preserving them continues a tradition in breeding and selecting that takes us back to our most distant ancestors – in the case of the runner bean, the first farmers of Mesoamerica some 6,000 years ago.

Some Like 'em Hot

Chilli is not so much a food as a state of mind.
Addictions to it are formed early in life and
victims never recover.

Attributed to Margaret Cousins (1878–1954)

I once lived opposite a pub called The George. The locals included a cohort of fellows who prided themselves on their ability to eat seriously hot chillies. I had a reputation for growing some of the finest and hottest chillies in the county. The types in my garden were legion – long, short, fat, thin, round, knobbly, red, brown, yellow. By mid-summer I would have new varieties for my drinking pals to critique, but why was it that the only thing that mattered to any of them was how hot my chillies were? Flavour, fruitiness, the subtleties of heat and spice, and in what dishes they would excel, were of no import. All they cared about was who could handle the hottest chillies without either being sick or seeking medical help. It was one named Lemon Drop that finally resolved my dilemma.

Bravado – At a Price

Lined up alongside the pints of ale were four small dishes, each holding a few very different chillies. First up was a delightful, mild, sweet and well-flavoured, pencil-thin, green variety (20cm/8in long) from Turkey called Sivri Kil Biber, which is grown principally to be eaten as a pickle. The tasters dismissed it contemptuously as being pointless. Alongside was a tiny round specimen, pale red in colour, pleasantly spicy and no larger than a pea, which I call Rodriguez Tiny because it is found only on the island of that name. A few raised eyebrows, gentle munching and polite comments were made, but no big deal…Third up was a scarlet cayenne type. Long, thin, slightly curved, fruity and spicy, measuring 15cm (6in) in length, I had discovered this growing on the Minahasa Peninsula of Sulawesi, one of Indonesia's largest islands, and also named after its place of origin. Often eaten fresh, it is, however, grown primarily for drying and to make into chilli powder. The nibblers commented approvingly but felt they weren't being sufficiently tested. Hadn't I something hotter?

Now was the turn of Lemon Drop. A handful of bright yellow, slightly ridged, conical chillies, approximately 5cm (2½in) long, lay enticingly in the bowl. I use this little darling with moderation in the kitchen. It has an intense citrus flavour that competes with the mouth-numbing heat and spice, which kicks in a few moments after taking the first bite. I knew what was coming, unlike my fellow drinkers. At first there was nonchalance, followed soon after by expletives. Then came much downing of beer to extinguish the agony – in vain! It was a pleasure to behold and, as I had hoped, proved to be the last time anyone asked to see just how hot my chillies were. The true joy of Lemon Drop, used in moderation and with the right dish, was lost on these gastronomic heathens, but it did the trick.

Where It All Began

Chilli peppers are members of the Solanaceae family and *Capsicum* genus – from the Latin *capsa* meaning 'box'. There are about 25 wild species, of which five have been domesticated. I grow them all. The archaeological record tells us that peppers, both mild and spicy, were first domesticated about 7,000 years ago in Mesoamerica, the area of greatest diversification still of the most widely cultivated pepper species *C. annuum*. Immature fruits start off green or purple, then ripen into many shades of yellow, orange, brown, purple and red. Varieties include sweet bell peppers and their hot cousins, cayenne, pimento, jalapeño and serrano. Most can be eaten both ripe and unripe as well as dried, depending on the requirements of the chef.

C. pubescens – the species name here meaning 'hairy' – is commonly called the rocotto pepper and is an important part of Bolivian cuisine. It is also the most decorative pepper I grow. It is happiest in cooler climates, which is why it has a permanent home in my greenhouse. If I am clever enough to keep a plant growing through several seasons, it takes on the appearance of a bonsai, with gnarled bark and a miniature tree-like appearance. The beautiful purple flowers give way to plum-like fruits that contain black seeds. It is truly delicious too – spicy and fruity. It shares its native homeland, the Bolivian Andes, with *C. baccatum*, a mild and fruity chilli with lovely creamy flowers; it too is important in Peruvian and Bolivian cuisine.

The fourth domesticated species is *C. frutescens*, which is thought of as a primitive species, not much differentiated from its wild parent. These colourful plants, with small, erect, red and bright yellow fruits, are often grown as ornamentals.

They are pungent and spicy and most frequently known as bird's eye. The most famous variety (which I write about later) is the key ingredient in Tabasco sauce. A core part of Asian cuisine, they have been adopted by many other food cultures. Varieties come in many shapes: the Bishop's Hat lives up to its name but is not especially hot. There is also a variety known as *Ají Limón* or 'Lemon Drop', but it is nothing like its hot namesake with which I had tormented my fellow drinkers.

That namesake is a member of the fifth species *C. chinense*. Although genetically synonymous with *C. frutescens*, the two species are treated by botanists as being separate.[1] Native to Central America, the Yucatan and the Caribbean, this species is generally very hot indeed and includes some of the best named chillies, such as the well-known habanero, the colourful Dragon's Breath and the Trinidad Moruga Scorpion. The popular Scotch Bonnet is another variety, although easily confused with similar-looking chillies like the Bishop's Hat.

It is easy to differentiate between wild and cultivated chillies. Domesticated capsicums – as they are botanically referred to – in all their myriad shapes and colours remain on the plant even when fully ripe and desiccated. All known wild species are deciduous perennials and have small, spicy, cherry-like fruits that fall to the ground when ripe and are much enjoyed by birds who, unlike humans are unaffected by the chemical capsaicin that gives chillies their heat. Their taste buds appear immune to the chemical, which is produced by the plant as a means of protection against herbivores.

Archaeological evidence suggests that the tribes of Mesoamerica were eating wild chillies at least 10,000 years ago, while the earliest finds of cultivated chillies were in a cave in the Tehuacán Valley in south-central Mexico and date from 5000–6000 BCE.[2] Wild chillies, like many other parents of domesticated crops, are weedy plants which do well in disturbed ground. This habit would have meant that they thrived alongside Neolithic hunter-gatherers whose daily activities around settlements would have encouraged them to flourish. Today, wild chillies are still foraged and planted by indigenous peoples in Mesoamerica and South America.[3]

A Surprise Discovery

On arriving in the Bahamas in 1492 Columbus was immediately struck by what the native Arawak people ate. On the menu were crops he had never seen before, including a very spicy fruit that was an essential ingredient of a dish kept continuously simmering on the fire. He called this stew a 'pepper pot' because the fruit had added a flavour not dissimilar to that of the black pepper, *Piper nigrum*, which grew only in southern India and that he had brought with him to enliven his meals on the voyage across the Atlantic. What the Arawak were growing was most probably a close relation of the variety from which we get Tabasco sauce. They also grew the classic slender cayenne type which, although hot, was sweeter. Archaeological evidence suggests that the Arawak people had brought chillies with them when they migrated from the mainland several hundred years earlier. Columbus recognised the importance of chillies and that these peppers were distinct from the completely different species, *Piper nigrum*, that grew in southern India.

The number one goal for Columbus was to find an easier way to trade pepper. Now he had stumbled on an alternative that could possibly be grown at home – unlike black pepper which required a tropical climate to flourish. The Arawak called the fruit *aji*, a name which was adopted by Columbus and everyone in Spain who embraced the amazing new spice. It remains the common name for many types grown across South America.

Columbus returned to Spain with his trove of new foods, leaving a small garrison that upset the locals and resulted in their stockade being bombarded with calabashes filled with ground-up *aji* mixed with ash. An eye-watering experience for the hapless defenders.[4]

During his second trip to the New World Columbus noted in his journal that the pepper the locals used as a spice was abundant and more valuable than either black pepper or its cheaper substitute, *Aframomum melegueta* (grains of paradise) peppers. The melegueta pepper – also known as the Guinea or Ginnie pepper – is, in fact, a member of the ginger family, Zingiberaceae, which is native to Africa's west coast and was used in Venice from the thirteenth century as a cheap substitute for black pepper. Soon after Columbus realised this, the Portuguese were dispersing chillies across their empire. These were cayenne types that they called Ginnie peppers. In Portugal's Brazilian colonies tiny local capsicums, possibly a bird's-eye type, were referred to as malaguetas by the African slaves who were transported from Guinea. All this made for a confusing situation.[5]

Spain was not the only global player at the end of the fifteenth century. Portugal was a great colonial power too. As well as occupying a number of Atlantic islands, large areas of Africa and South America, and outposts in South-East Asia, it also had colonies in Goa and Calcutta. Portugal and Spain were on reasonable terms as the fifteenth century drew to a close, although there existed a trade exclusion agreement between them. However, the Spanish had few ships; in fact, transatlantic activity between Spain and the New World was fairly limited. So, it was the Portuguese who were responsible for the chilli spreading around the world in double-quick time. The first that started their global takeover were principally cayenne types, although bird's eye and naga chillies, which are varieties of *C. chinense*, soon found appreciative new homes too.

When Cortez arrived in Mexico in 1519, he found plenty of different peppers which the Aztecs in their Nahuatl language called *chili*. Hence 'chilli pepper', the name this crop has retained globally ever since. The first botanists to classify peppers grouped them all as one species. As with so many other vegetable discoveries, names could be contradictory and confusing, made more so thanks to the pepper's lightning-fast colonisation of the world.

The first illustrations of chillies were in an herbal by the German botanist and physician Leonhart Fuchs (1501–1566), published in 1542, which is proof that the chilli was known in Central Europe within a half century of Columbus finding it and that distribution of the chilli began well before Cortés's conquest of Mexico some 20 years earlier.[6] The Scottish physician George Watt (1851–1930), who worked as an economic botanist for the British in India towards the end of the nineteenth century, suggested that the Portuguese were exporting chillies from India and competing with the existing trade in black pepper. Thanks to the original spice trade which made use of the sea lanes of the Indian Ocean and the Arabian Gulf as well as land routes from India to Europe, there was a complex and sophisticated communications network in place that ensured new foods brought first to the Portuguese colonies could, within a few short years, be grown and assimilated into all the cuisines of the known world.[7]

The rapidity with which the capsicum spread to Africa and Asia confused many a botanist. Nearly 300 years after this culinary colonisation, the Dutch botanist Nikolaus Joseph von Jacquin (1727–1817) named a species of capsicum, *C. sinense*, that he had come across while plant-hunting in the Caribbean in 1776 because he thought it had originated in the Orient, so ubiquitous a global crop had the chilli become by then. In fact, the species was only officially classified in 1957, with the new spelling, *C. chinense*, presumably to differentiate it from the incorrect interpretation made by Jacquin.[8]

Perhaps surprisingly, chillies did not accompany the other Mesoamerican crops, maize, squash and beans, into North America until the arrival of the Spanish in what is now New Mexico at the very end of the sixteenth century. The arid environment of the Southwestern States made growing the earliest domesticated chillies difficult, although Native Americans were consummate irrigators centuries before the arrival of the Spanish. There is a wild species of chilli *C. annum* var. *glabriusculum*, native to the deserts of northern Mexico and the US, which is foraged by the Tohono O'odham people of southern Arizona. Highly valued and delicious, it is called *chiltepín*.

An Indian Discovery

As peppers spread around the world, they acquired their own unique place in local cuisine with names that are a familiar part of our vocabulary. Most successful of all species was *C. annuum*. The Indians fell in love with it and today practically every region of the sub-continent has its own special variety. This became very evident to me when, early in 2019, I was travelling across Rajasthan, curious as to how well *desi* – local – food crops were flourishing. I had been told by the head of the Agricultural Research Station in Bikaner, Dr P.S. Shekhawat, that one culturally important plant I had heard a great deal about, the Mathania chilli, was effectively extinct due to the introduction of modern cultivars.

Floral and aromatic when first cut, the Mathania chilli was highly valued for its fine culinary qualities. It is a large, slender fruit with more spice than bald heat. A staple of Rajasthani cuisine for generations, it was an important crop for those few farmers who were able to grow the chillies in semi-arid conditions. However, what is now being sold is but a pale imitation of the original fruit. On my travels everyone I spoke to about this famous chilli grumbled about it having changed in the last 200 years or so. Due to problems with irrigation and small crops, farmers have been switching from their own saved seed to F1 hybrids, which give greater yields but, perversely, require more water and are more sensitive to low nighttime temperatures, making them less robust in extremes of climate. Despite being self-fertile, chillies will cross-pollinate. So, the introduction of new varieties into a farming system where the average holding is about 20 acres (8 hectares) has meant that the Mathania chilli now consists mostly of introduced genetic material. Sellers in local markets and every farmer I spoke to about this chilli all said the same thing: the genuine Mathania was lost. Could this really be true? If so, it was a story that had become familiar to me over many years of tracking down rare and endangered local varieties. That's why I found myself, after a half-hour bumpy ride in the back of an ancient jeep, a little shaken up but no less curious for that, at the gate of a small compound about 25 miles north of Jodhpur. The gently undulating countryside near the town of Mathania is at the centre of a region once famous throughout India for its prized chillies.

The compound belonged to the family of Mrs Devi. It was a corral of round, white-painted cob rooms set in an immaculate earth courtyard. We ate chapatis made from her own landrace millet and a curry of yoghurt from the milk of her heavenly Rajasthani sheep, flavoured with seeds from a nearby sacred desert tree *Prosopis cineraria*, known locally as Khejri. There in a corner of the compound was a small pile of chillies, harvested from a tiny plot where a dozen or so plants grew. My guide Pritam Singh, also a farmer, was beside himself. He needed just one sniff, one bite and, with a smile that would have lit up the darkest night, declared that this chilli tasted and smelled of his childhood. This was the real thing. I too sniffed and ate. Scrumptious. Mrs Devi kindly gave me a small handful of seeds from this latest crop. She had been growing her chillies for as long as she could remember, as had her mother before her. There were no other chillies being grown for miles around. I believed I had stumbled across something all the experts had told me was extinct – a genuine landrace Mathania chilli. Of course, I am now growing them and have returned some seeds to Mrs Devi and plant material to the scientists in Bikaner to do some genetic testing to see if I am right. For the record, my first crop was prolific and many a meal enlivened by them.

The Mathania chilli is a cayenne-type. Wherever in the world I find myself, I come across these long, slender fruits. They may share similar culinary qualities, but for the communities that grow them they are all unique. So how did this type of pepper get its name? Cayenne is thought to be a corruption of the name *quiínia*, which comes from the now-extinct language of the Tupi people who inhabited parts of what is now French Guiana at the time of Portuguese and Spanish colonisation in the sixteenth century. The capital city of that country is Cayenne. The seventeenth-century herbalist Nicholas Culpeper (1616–1654) describes this pepper as being synonymous with the Guinea pepper. Confusing indeed, and completely wrong. In fact, the name Guinea pepper was also used until the nineteenth century to describe bird's eye chillies or, as the Portuguese call them, piri-piri. Historical evidence suggests that, from the start, the Portuguese distributed all three chilli species, *C. annuum*, *C. frutescens* and *C. chinense*, among their colonies. However, cultural preferences meant that the cayenne type was dominant in India. The much hotter bird's eye was preferred among the Malay tribes of Indonesia and the Philippines and across the Far East.

A Bird's Eye Bonanza

Finding truly local capsicums is one of the great joys in my life and there are over a dozen bird's eye chillies in my collection that come from around the world. This type of chilli is grown across much of Africa, where it was first introduced to the east coast settlements by the Portuguese over 500 years ago. *Piri-piri* in the Mozambique Ronga language means 'very hot pepper'. In central and east Africa, the Swahili name is *peri-peri*. Piri-piri sauce and the delicious Portuguese national dish piri-piri chicken have their origins in Portugal's colonies in southern Africa, especially Mozambique. With regional tweaks, the sauce is ubiquitous throughout southern Africa. This little chilli is cultivated on a large commercial scale from tropical central Africa to the temperate south. I have found it growing wild in the Masai Mara in Kenya and as an escapee on verges and abandoned ground across Nigeria. However, nothing beats the encounters I have had with this little beauty further east.

While in Singapore, I discovered a kindred spirit, Kenneth Liang. Although we met to discuss a co-production deal, we very quickly realised that we had a shared passion for growing stuff and Kenneth's small garden was home to a bird's eye chilli that had been in his family for generations. Needless to say, this very hot example is now in my library. The Portuguese, who conquered Malacca – just up the coast from Singapore – in 1511, also brought the same chillies they had started growing in their African colonies to this part of South-East Asia. Perhaps Kenneth's chilli is a direct descendant?

Bird's eye chillies were brought to the islands of the Indian Ocean by the Portuguese too. In 2007 I was making a film about the very last surviving example of a plant endemic to the tiny and idyllic island of Rodrigues, *Ramosmania rodriguesii*, which is known locally as Café Marron. Once filming was over, it was time to check out the market of the island's capital, Port Mathurin. There, in large quantities, were jars of tiny, native, pickled, green bird's eyes. But how did they get there? The island was first made known to the Old World in 1509 by the Portuguese explorer Diogo Fernandes Pereira while en route to Goa. The island got its name, however, from another Portuguese sailor, Diogo Rodrigues, who spent some time on the island in 1528 while sailing home from Goa. The romantic in me likes to hope and believe that a few seeds were left behind by one or other of the Diogos.

I have had other more fiery encounters with the bird's eye, most memorably in Malaysia. I dared to test my ability to eat this very hot chilli, known in Malay as *Cabe Rawit*, which is the principal ingredient of a condiment, chilli padi, that is designed to spice up any dish. It is found on practically every restaurant table throughout South-East Asia. Probably on account of already having drunk rather too much of the local beer, I bet my fellow diners that I could eat not just one bowl, but ten. Well, the first portion was the hardest to consume, but with a numb mouth and stomach of cast-iron, I steadily ploughed my way through each bowl as it was brought to the table by an incredulous waiter. What I didn't know at the time was that the staff had opened a book on me too and, among fellow diners and waiters, increasing amounts of cash were now at stake. I am proud to relate that I won the bet, but my alimentary canal paid a terrible price – I still feel shame at indulging in chilli machismo while under the influence. It shall never happen again.

I have stumbled across hedges of bird's eye chillies in Myanmar and, while travelling through French Polynesia, a small plantation of tiny and fruity, very hot examples. However, it is a handful of deep crimson beauties, 2.5cm (1in) long, and given to me by a dear old lady in a market in Tavira, southern Portugal, that mean the most to me. They don't taste so different, but they are a genuine piri-piri heirloom, a direct descendant of the very first chillies to arrive in Portugal in 1492.

The Ultimate Condiment

The most famous bird's eye chilli of them all surely has to be the Tabasco pepper and it is a very close relative of those small ones that Columbus found on his arrival in the New World. Tabasco, on the Gulf of Mexico, is one of the country's southernmost regions, right in the centre of Mesoamerica. The pepper is native to the area, which is just a short journey by canoe to the islands Columbus first saw. So, it is little surprise that the Arawak tribes, who lived in southern Mexico and had colonised the islands, grew this particular chilli in their gardens. Also within easy trading distance for the Arawaks across the Gulf of Mexico lies Avery Island (not actually an island but part of the fertile coastal region south of Lafayette). In 1868, a local resident, Edmund McIlhenny, had an idea. A keen gardener and food lover, he wanted to pep up the somewhat bland local cuisine, so he sowed some seeds of a particularly juicy little chilli that had come across the Gulf from southern Mexico. He so liked the fruit that he founded the McIlhenny Company to produce a condiment that has become the best-known chilli sauce in the world.

The Tabasco pepper is closely related to its wild parent and so has many genetic links to another member of the Solanaceae family, tobacco. Sadly, it is particularly vulnerable to a nasty pathogen, tobacco etch virus (TEV), which was a big problem for the McIlhenny Company in the 1960s. It took nearly a decade before Auburn University in Alabama developed a new cultivar, the Greenleaf Tabasco pepper, in the early 1970s. The breeders were thus able to preserve all the traits that gave the native Tabasco pepper its unique qualities, but now with considerable resistance to TEV. Tabasco sauce was saved for the world! Thankfully, the original pepper is still widely grown across the southern United States and in Mexico, while its wild parent continues to flourish as a weed.

A Matter of Heat

Chillies, regardless of species, have varying degrees of heat, fruitiness and spiciness. Heat can be measured on the Scoville scale.[9] William Scoville (1865–1942) was a pharmacologist who in 1912 invented a method whereby he would extract the chemical capsaicin – the stuff that makes chillies hot – and dilute it in water until, using a panel of five tasters, the sensation of heat is no longer present. The key genetic difference between sweet and hot peppers is that the former has a single recessive mutation which blocks the production of capsaicin. Hybrid sweet bell peppers can have an average Scoville Heat Unit (SHU) of zero. The bird's eye chilli I suffered from eating to excess in Malaysia would have had a SHU of between 50,000 and 100,000. One cubic centimetre of the extract would have needed to be diluted with 50–100 litres (11–22 gallons) of water for the heat to have become undetectable to my taste buds.

There are growers around the world who obsess over breeding ever hotter chillies. The most successful can be found in the U.K. and the USA. The first British one to make it into the *Guinness Book of Records* was the Dorset Naga, bred by a very talented couple of traditional plant breeders, Joy and Michael Michaud, who run Sea Spring Seeds in South West England. In 2006, their chilli scored 923,000+ SHU, making it the world's hottest chilli. The Michauds' claim today that their Dorset Naga scores at 1,221,000 SHU. I dare to suggest that chilli growers who claim huge SHU numbers and fishermen with the size of 'the one that got away' share much in common.

The task of creating hotter and hotter chillies goes on, though for what purpose other than to reach an ever-higher number I know not. Eating chillies that are this hot is something I studiously avoid. The current world-record holder is the Carolina Reaper, grown by Ed Currie of the Puckerbutt (I'm not kidding) Pepper Company. When tested in 2017, it checked out at an average of 1,641,183 SHU, leaving even the latest Dorset Naga trailing some way behind. Every year grown men and women – they call themselves chilli heads – compete to see who can eat the most Carolina Reaper in one minute without being sick or collapsing. The record achieved in 2017 of 120g (4oz) remains, at the time of writing, unbeaten.[10]

Naturally Hot

Champion chillies are nearly all varieties of *C. chinense*, although the Carolina Reaper is the result of a deliberate cross between *C. chinense* and *C. frutescens*. Such crossing has also been occurring in an unplanned way for centuries. Before the craze for breeding super-hot chillies really took off about 15 years ago, it was generally considered that the world's hottest came from northeast India. There are countless landraces grown in this region, all very similar, with names such as Naga Jolokia, Bhut Jolokia (with an SHU of one million) and Bih Jolokia (*Jolokia* is the Assamese word for 'capsicum').

I have on occasion happened upon some particularly fiery examples, most memorably in neighbouring Myanmar. The country's cuisine varies considerably from region to region. However, I have never associated it with extremely spicy food. So, imagine my surprise when one morning in Yangon, while wandering through a street market at dawn, I came across another wonderful member of the 'Granny' tribe selling vegetables from her garden, including radiant red chillies. The fruit were about two inches long with a knobbly, almost rough skin, quite plump but narrowing towards the tip. They were redolent of the Naga types I was familiar with. I sniffed. The fruit was quite fruity, and when I pricked the surface with my thumbnail and tasted the juice it was intensely spicy. I had to buy some. However, the neighbouring stallholder was determined that his neighbour should, quite rightfully, properly profit from the foreigner at her stall. I paid what he suggested and like to think I was her best customer that morning. I was certainly the most appreciative.

On getting the seeds home, the chillies performed very happily, producing an abundant crop of the most blisteringly hot specimens I have ever grown. Considering that northeastern India borders Myanmar, I guess I should not have been surprised to discover the chilli I named Burmese Naga shares the same fiery qualities as its sibling *Jolokia*. Half a chilli in a curry is more than enough, although it is a valuable ingredient in seriously hot chilli sauce. Perhaps I should see if my old pals at The George fancy trying them.

The Sweet Ones

The sweet bell peppers that we see in glorious technicolour on supermarket shelves are mostly hybrid cultivars; they make bland and uninspiring culinary fare. Yet, traditional large, mild peppers, which are their progenitors, have been an established part of the food culture of the Middle East and North Africa for centuries. From those first introductions have come the *pimiento* type: a key part of Spanish cuisine. The smaller, round and multi-lobed varieties, like the one from the Ukraine I describe in my introduction, are the foundation of many Turkish and East European dishes. Sweet peppers were favoured by the Spanish and Italians from the middle of the sixteenth century and have evolved through several centuries of cultivation into a vast number of local varieties.

Nations and cultures are justifiably proud of their local varieties, many of which have Designation of Origin Protection (DOP) status under EU law. The Padrón pepper is one. Mild but known for the occasional hot ones in a crop, this variety is cooked when green and unripe. Although now grown in the USA, the real thing can only be sourced from its original home in the municipality of Padrón in Galicia, northern Spain. Most famous of all Spanish condiments are the sweet and hot *pimentón*. A type of *C. annuum*, they are considered by some to be another form of paprika, which is a major part of Hungarian cuisine. However, the Spanish, with considerable justification and history on their side, claim their version is unique. The story goes that Columbus, on his return to Spain from his second trip to the New World in 1494, gave the *pimentón* to Queen Isabella and King Ferdinand, who considered it rather too hot for their taste buds. However, monks from Guadalupe thought otherwise and they shared it with other monasteries across Spain. Today the spice is grown in two regions: Murcia on the southeast coast, and the more famously named *La Vera*, from the Cáceres region of Extremadura in central Spain, where Columbus reportedly gave the king and queen their first taste.

Paprika, which is Hungarian for 'pepper', on the other hand, would have been the product of dried chillies of the pimento type that found their way into Hungary from the east. The Turks introduced ground chilli to the Balkans during Ottoman rule in the eighteenth century. Before then, the pepper was used as a treatment against typhoid and much valued as a decorative houseplant. Until the 1920s paprika was a hot, ground spice. Then a breeder in Szeged, southern Hungary, apparently came across a plant bearing milder and sweeter fruit, which became the principal type used for today's paprika.

Another truly wonderful type of *pimentón*, which has DOP status, comes from the Basque region of southwest France. Consumed fresh as *Gorria*, which is Basque for 'red', when dried and coarsely ground it becomes the revered condiment, Piment d'Espelette. It is a fiercely protected crop which is only grown commercially in that region. Getting hold of seed can be a challenge, and I was fortunate to have a French friend and fellow chilli-lover who had been given some seeds which he shared with me. I rate this hot and sweet chilli pepper very highly indeed and always have a supply of coarsely ground flakes in my larder.

I found another sweet and spicy variety while on holiday in Morocco exploring the giant sand dunes in the east of the country, close to the Algerian border. Happening across an oasis resplendent with palms that surrounded small, irrigated fields of alfalfa, maize and cash crops of tomatoes and herbs, I spied some three-lobed, conical, deep scarlet peppers, one of which could easily fit in the palm of my hand. The farmer told me he had been growing them all his life and that they were the true Moroccan desert pepper. They have a wonderful fruity sweetness and complexity, with sufficient heat to enliven a goat tagine. Dried and coarsely ground, they make a superb condiment. Ubiquitous in Morocco, this pepper is synonymous with one of the most famous of all: the Aleppo pepper.

Aleppo was a great centre of plant breeding and agriculture for centuries until the civil war in 2011. While in the city that year doing my usual market trawls, I visited a number of seed shops and bought several varieties of sweet, hot peppers, including the famous Baladi or Aleppo pepper. Ripe fruit is de-seeded and dried, often with the addition of salt and oil to add depth to the redness of the final product. The result is a coarse powder that is part of much Syrian cuisine.

I cannot imagine life without peppers, in all their glorious forms. Barely a day passes when I do not include them in my cooking. Growing them gives me enormous joy. I always give them a cheery 'morning all' on opening the greenhouse door soon after sunrise and they are a constant topic of conversation with fellow growers; packets of seeds from my collection are hoovered up by fellow aficionados when I attend seed swaps. Every time I harvest some to take to the kitchen, I am reminded of their provenance, how important they are to countless cultures, and of the many wonderful adventures I have had discovering them.

Not Just for Hallowe'en

As soon as the squash begins to run its arms
across the ground it is ready for the kitchen
Antonio Francesco Doni (1513–1574) –
La Zucca (*c*.1541)

Among the very small number of overrated and abused vegetables I have had the misfortune to eat and occasionally grow, it is the vegetable marrow that takes top spot. Were my mother still alive she would now take great offence. Her stuffed marrow was a meal I looked forward to with utter dread – watery and tasteless, the greying flesh redolent of ancient underwear and encased in a once colourful skin that was home to a stuffing of chewy mince and undercooked onion, devoid of seasoning and purged of flavour by time spent too long in the oven.

What's in a Name?

The vegetable marrow, a quintessentially British crop, is one of eight distinct types of edible fruits of the species *Cucurbita pepo*, collectively known as summer squash (*Cucurbit* is Latin for 'cup or flask'). The botanist Harry S. Paris has spent much of the last 30 years studying them, breeding new varieties, and coming up with his list of eight types[1] which are: marrow, pumpkin, scallop, acorn, zucchini, cocozelle, crookneck and straightneck – the last two were only discovered by botanists 150 years ago, being grown by native North American Indians. As we will see, naming family members has been fraught with confusion over many centuries. Was my mother's giant marrow actually an overgrown courgette (as the British and French call zucchini), and when is a zucchini really a zucchini? The word courgette is a diminutive of the French word *courge*, meaning 'gourd'. And although interchangeable with zucchini, it is also frequently used by seed companies to describe other types of summer squash. Zucchini is the plural diminutive of the Italian for gourds and squash, *zucca*. Do all these distinctions matter when all one really needs to know is if it tastes good. Humour me please, dear reader; to this particular vegetable anorak, it does!

Known as 'summer squash' because they mature early and are not generally stored for eating through the winter, patty pans or scallop courgette, straightneck, crookneck and cocozelle all describe shapes of fruits that we seldom think of as types of courgette. Seed companies rarely, if ever, make a distinction, and most of the descriptions of 'zucchini' or 'courgette' on seed packets include these types.

The true zucchini is, in fact, a very recent development. First described by Domenico Tamaro in his book *Orticultura* in 1901, he lists a type of fast-maturing, bushy *zucca* which produces long, cylindrical, dark green fruit flecked with white, best eaten when young. The word zucchini had been used to describe all manner of immature summer squash in the preceding centuries by Italian chefs and gardeners; all botanically incorrect. Nothing much has changed. Examples include traditional 'courgettes' with Italian and French names such as *Striato d'Italia*, which is a variety of cocozelle, and

Ronde de Nice, a member of the pumpkin group. Claimed by many seed sellers as a French 'heirloom', it is remarkably similar to *Tondo di Nizza*, which comes from Turin! What is clear is that the immature fruits of many different types have been consumed with much relish by Italians for more 500 years. Today, countless named varieties of summer squash in all their shapes and colours are loved and loathed in equal measure. The giant cylindrical specimens which my mother took such pleasure in force-feeding her children, and that continue to grace benches at British horticultural shows, may have been the result of careful selection by British breeders 200 years ago but today are often no more than 'courgettes' that have been left to mature.

I used to find summer squash rather dull, but that all changed when I returned from a seed-hunting trip to Syria with a packet of a locally bred variety. With a much higher fibre content than your average type, it had flavour too! Today I grow nothing else and proselytise endlessly about my 'Syrian courgette'. Unimpeachably yummy when harvested at no more than 12cm (5in) in length, it is slightly pear-shaped, which, according to Paris's description, means it is a marrow type and if allowed to grow will become the much despised – by me at least – vegetable marrow. Trouble lies ahead for the inattentive gardener when growing this particular crop. Blink and the thing has doubled in size. All I need is to be absent from the garden for a day or two and those glorious, green, speckled fruits are now only good for soup, the compost heap or the marrow class at a flower and produce show. The giant vegetable marrow may well be a cornerstone of traditional British cuisine, but it is fast losing any residual allure as a worthy crop. The immature fruitlets, picked when just a few days old and still able to fit in your palm, are a different story.

Four Types of Squash

The story of the domestication and development of all types of squash started 10,000 years ago in the New World. However, the naming of this vegetable has a long and convoluted history, the result of human error. The name squash derives from the Native American Algonquin *askoot asquash*, which translates as 'eaten raw'. Raw courgette makes a great salad. Today squash is used to describe four species and quite arbitrarily interchanged with another name, pumpkin, which derives from the Greek *pepon* and the Latin *pepo*. *Pepon* was first used by the Greek physician Galen (129–*c*.216 CE) to describe ripe cucumber. I get especially irritated when I see translations of recipes that date back to Roman times, where squash is used to describe a fruit that it most definitely is not! It was the first British colonists arriving in New England – Algonquin territory – early in the seventeenth century who anglicised the name to squash.

As we have seen, the species *C. pepo* includes a type of summer squash called pumpkin. However, the p-word is also used to describe some members of the other three species of domesticated squash, *C. maxima*, *C. mixta* and *C. moschata*. In this chapter, I use 'squash' as a generic description for most types of New World species of cucurbita. The debate over when a squash is a pumpkin or a pumpkin is a squash manages to confuse not only the lay reader but enthusiasts like me, and there is still no internationally accepted definition for either. The popular naming of a type has a more meaningful cultural relevance than a botanical one – which is unambiguous – and becomes evident as we follow the journey of the domestication of the four species that are central to my culinary pleasures.

So, just how important are squash to our collective food cultures? Yet again we have Christopher Columbus to thank for introducing the squash to the Old World, when he brought it back from his expedition in 1492. Writing on the history of squash in *Vegetables of New York* (published in 1928), Professor G.P. Van Eseltine said, 'The history of the cultivated cucurbita, if written in full, would form a large part of the story of the development of agriculture in the tropics and sub-tropics of both Old World and New.'[2]

Prior to Columbus's voyages to the New World, the bottle gourd, *Lagenaria siceraria*, had been widely grown in the Old World, used both as a receptacle and for its seeds, which are highly nutritious and make excellent oil. Its flesh was occasionally eaten but was far inferior to the new arrivals from the Americas. However, sixteenth-century botanists confused gourds with squash and made matters worse through incorrect classification and naming that was only resolved some 200 years later.

The earliest archaeological records of indigenous American tribes show that they were using both gourds and squash primarily as receptacles but also for their nutritious seeds. Squash and gourds come in a dazzling variety of colours, shapes and textures, thanks to their ability to freely interbreed. However, both species evolved independently and cannot cross with each other. Neolithic farmers of the New World were able to select from the indigenous wild squash traits that gave them greater size, bigger seeds and, most significantly, sweeter and less fibrous flesh. They either never bothered with further improving the bottle gourd or else it was less responsive to domestication syndrome. This describes a process by which wild plants acquire traits that make them worth cultivating – for example, loss in part or whole of seed dormancy, which means a crop will have more uniform germination and fruit ripening. An important element in this first phase of domestication is the loss or reduction in chemical defences that can cause sickness or death. The second phase of domestication, key in the case of squash, exploits these traits to select for increased size of fruit, colour, shape and edible seeds as well as something we value most highly: flavoursome flesh. Although the juice can be highly toxic, bottle gourds continue to be grown in the New World for their seeds but also as containers and ornaments: definitely not their flesh.

Pompeons, Melons and Gourds

The story of the domestication of squash begins with *C. pepo*. It is native to North America and has been cultivated by indigenous peoples for thousands of years. The wild parents of *C. pepo* were similar in many ways to the native African gourds. Small, with very hard skins, a bitter and fibrous flesh, and few seeds. In fact, many so-called 'ornamental gourds' which continue to be popular among some growers today are little different genetically to wild *C. pepo*. The bitterness is due to the presence of the chemical cucurbitacin, which the plant employs as a defence against herbivores. Even small amounts can cause stomach ulcers and has been known to kill unsuspecting eaters.

The lengthy domestication of *C. pepo* resulted in two sub-species. The first, *C. pepo* subsp. *pepo*, was brought back by Columbus on his return from Hispaniola in 1493 and begat the eight different types of summer squash described earlier. The second, *C. pepo* subsp. *ovifera*, includes the storing type of acorn squash, which is a regular in my garden. Dark

257

green, almost black, with deeply ribbed fruits, Table Queen is a favourite. It was bred by the Iowa Seed Company of Des Moines and first sold in the USA in 1913. It is described in *Vegetables of New York*[3] as being the same but tastier than one grown by the Arikara tribe of North Dakota who, as consummate horticulturalists, had no doubt been growing it for centuries before a plant breeder 'improved' it. Another member of this sub-species is the bright orange specimens generally called a pumpkin that are the decorative centrepiece of Hallowe'en celebrations and the key ingredient of 'pumpkin pie'. In the sixteenth century there was no systematic means of classification, so all cylindrical varieties were known as marrows and crooknecks, and round types were called pumpkins, scallops, melons or acorns.

Fifty years after the first New World squash arrived in Italy, many botanists used both pepo and pepon to describe types of melon – which are native to the Old World – too. By the second half of the sixteenth century, however, melon was also used in English to describe pumpkins! *Courge* – which was subsequently anglicised to 'courgette' – was also used by explorers and botanists alike over the following 200 years to describe pumpkins, squash and bottle gourds. Although botanists were already recognising distinct differences and traits in the various groups and types of cucurbits, they persisted in calling all forms of squash pompeons, melons and gourds. Even Carl Linnaeus (1707–1778) considered all gourds and squash to be of the genus *Cucurbita*. Gourds were only assigned their own genus *Lagenaria* by the botanist Juan Ignacio Molina (1740–1829), after Linnaeus died. The American botanist Edward Lewis Sturtevant (1842–1898)[4] suggests that the different shapes of both summer and winter squash led to the confusion in their naming. Large, round ones were called pompeons; those with hard rinds kept for winter use were known as gourds and small round ones as melons. One of the easiest ways to distinguish gourds from squash is that the former have white flowers and the latter, yellow, but it took over 200 years for the penny to drop. All four New World squash also have easily identifiable and distinct appearances.*

After a slow start – like so many vegetables at the time, considered only fit for the poor – squash became a popular food in European diets. By the end of the sixteenth century the British had lumped together all the different species of large, hard-skinned squash that had been brought back from their American colonies and called them 'pumpeons'. They grew well in an English climate and soon the name morphed into pumpkin. Combining the sweet flesh with dried fruits, apples and spices, and then baking this in a pastry base was a popular dish. Thus was born what was to become an icon of colonial American cuisine: pumpkin pie. This love affair with hard-skinned types of *C. pepo* and other hard-skinned species, mostly *C. maxima* (which is what the Algonquin cultivated), coincided with the colonisation of North America. Their great value in the USA in the seventeenth century was

---

* *Cucurbita pepo* has deeply cut leaves with spiny hairs and five-ribbed fruit stalks which do not swell at the base of the fruit as they mature. The skin becomes very hard when ripe. *C. maxima* has large, round, undivided leaves with stiff hairs that are not spiny. The fruit stalk is round and unribbed and can often be of much larger diameter than the stem. *C. moschata* has dark green, blotched with silver/white, angular, undivided leaves which are hairy but without spines. *C. mixta* have leaves similar to those of *C. moschata* but they are more jagged and have white flecks.

as animal fodder and, *in extremis*, for human consumption. The early colonisers did not have ovens, so would cut the top off their pumpkins, remove the seeds and fill the cavity with a mixture of milk, spices and something sweet like honey. After replacing the top, the stuffed vegetable was pushed into the ashes of a fire to bake slowly. What resulted was a thick creamy custard, not dissimilar to the filling of today's pumpkin pie. The flesh was also used to make a type of bread or cake and fermented to make beer. As Europeans moved west, they were able to trade and cultivate squash, which formed the staple of much Native American food culture.

Made in the USA

Only *C. pepo* is a summer squash. This means that it does not keep for long periods, unlike all the other varieties which are known as winter squash. There are two types of the species *C. moschata*, which have good keeping properties like other winter squash, and these also have an important place in American food culture. One, popular in the US but less well known beyond, is the giant crookneck. The Seminole tribe that first settled in Florida early in the eighteenth century grew these giant crooknecks (which should not be confused with the crookneck courgette, which they called *Cushaw*). This squash was classified as a separate species, *C. mixta*, by the botanist K.J. Pangalo in 1930, although not fully described until 1950.[5] Until then it was considered a variety of *C. moschata*. In all honesty, I must admit that I will need some persuading to grow this species again as it is not as flavoursome as most other crookneck squash. It is better appreciated for its large, delicious seeds, which are a popular snack in Mexico and Guatemala.

The other type of *C. moschata* is butternut squash which is now catching up with the courgette in becoming a global food phenomenon. I grow them and love them. *C. moschata* is native to the humid lowlands of much of Mesoamerica. Although no wild relative has yet been found, this type of squash was being cultivated on irrigated farms at least 4,000 years ago. The oldest archaeological records found in Huaca Prieta, the coastal desert region of northern Peru, date from this time. It would seem that indigenous tribes of the region were growing squash before they grew maize. It is also evident from later archaeological finds that the fruit must have been of great cultural significance because the Moche potters of the region produced life-size reproductions between the second and eighth centuries CE. *C. moschata* was already being widely grown in the northeastern coastal regions of Mexico 3,500 years ago, and there is evidence that it was also part of the food culture of the Pueblo regions of southwestern USA before European colonisation.[6] With the arrival of the Spanish in the sixteenth century, it didn't take long for a diverse number of varieties to each become the number one winter squash throughout their West Indian colonies and also in Florida. By the end of the seventeenth century, they were widely grown in New England, known as – you've guessed it – pumpkins! Today, the majority of tinned pumpkin and commercial pumpkin pie is made from varieties of *C. moschata* – butternut or crookneck squash – with the true pumpkin, *C. pepo*, as Sauer writes '…left uneaten as jack-o-lanterns'.

Of Butternuts and Crooknecks

Hugely diverse in its centres of origin, of the two types of *C. moschata* it is the butternut squash which has been the subject of much modern breeding in the last century and dominates squash production in the USA today. Some giant crookneck cultivars are grown more for their appearance than – in my opinion – for flavour, such as Pennsylvania Dutch Crookneck which I grow because it looks magnificent even if it is not that flavoursome. It is a superb 'keeper' and, along with some of the earliest known American breeds like Canadian Crookneck, which was first sold by seed merchant Charles Mason Hovey in 1834, became an important part of an American's winter diet. Because all types of *C. moschata* like to grow in hot and humid conditions, the British climate does not favour them. However, modern F1 hybrid cultivars of butternut squash litter the aisles of supermarkets and plenty of seed catalogues list several with excellent culinary qualities which have been bred to flourish in cooler climates. However, F1 hybrid squash, tasty as many are, do not grow back true to type from saved seed, so I don't grow them. Fortunately, my favourite butternut, Waltham, is the result of some traditional breeding in the 1960s in Massachusetts, USA, so I am able to save seed – no problem!

Taking Over the World

The fourth species of squash I grow, and by far the most delicious, are winter squashes of the species *C. maxima*. Its wild parent is believed to be a weedy pioneer, *C. andreana*, which is native to the Río de la Plata region of Argentina and Uruguay. The social and cultural importance of this domesticated species dates back to its earliest use in pre-Hispanic Argentina, where it was one of the key crops of the Guaraní Indians who lived in that region.[7] Whole fruits, some 1,500 years old, have been found in Salta, a mountainous province of northwest Argentina, and later finds have been made in Peru. Domestication was restricted to the temperate regions of South America. The other three species were being grown right across the North American continent by the time the Spanish arrived early in the sixteenth century. They brought with them *C. maxima* which grows well in cooler climates, unlike other species of squash, and it quickly embedded itself in the cuisine of Native Americans. Soon there wasn't a tribe that didn't include at least one species in their diet, enjoying the fruit, the flowers and the seeds. By the end of the sixteenth century squash in many forms was ubiquitous in European colonies across North America.

The earliest varieties of *C. maxima* came in many shapes and sizes. Some were eaten like a courgette, as immature summer squash. Most, if stored well when ripe, could keep for six months or longer. Others grew to an enormous size. In the sixteenth century, early Spanish explorers, when travelling in the headwaters of the Amazon in Bolivia, were said to have found squash so large that one was almost too heavy for a man to shift on his own.

The most decorative and among the very tastiest squash I grow is Turk's Turban, which describes it perfectly. Its orange and red skin, fruit the size of a human head with a protruding 'button' on the top, often streaked with green, make it one of the loveliest – or ugliest according to your taste – of all squash. It was described by the naturalist J. Molina in 1782 as as a 'spheroidal fruit with a large nipple at the end, the pulp sweet and tasting not unlike the sweet potato'.[8] In the middle of the nineteenth century some taxonomists suggested that this magnificent squash should receive its own classification, *C. maxima Turbinaformis*. Sadly, in my humble opinion, this suggestion was never adopted. If I grew no other, my wife would be happy. Even so, every year, as with other species and in order to avoid cross-pollination, I grow only one variety of *C. maxima*, Turk's Turban, which takes the top spot.

As well as Turk's Turban, a very colourfully named variety, which is held in the Heritage Seed Library and is quite high on my list of good squash, is Chicago Warted Hubbard, an ex-commercial cultivar. It is a deep green, knobbly and bulbous beast from Illinois that was first sold there in 1894. Its name tells you that it is a hybrid cross between two different types of *C. maxima*. Hubbard squash can be the size of a football and shaped a little like a Chinese lantern. One of the most popular squash for the nearly 200 years, it was named after a Mrs Elizabeth Hubbard from Marblehead, Massachusetts, who first grew it in the 1830s.

I have US favourites of *C. maxima*, including Oregon Homestead Sweetmeat, an unclassified type which perfectly describes not only its provenance but also its flavour. *C. maxima* was the last of the squash family to be embraced by Italians and there are but a handful of local cultivars, of which *Marina di Chioggia*, from the Lombardy region in the Po Valley, is my favourite. Spectacular-looking, football-sized, warty and dark green, it conceals a deep orange flesh that makes for wonderful eating. Japan's squash heritage includes many highly distinctive cultivars that were bred from native South American types of *C. maxima*, which arrived from the New World in the eighteenth century. In fact, some of the most extraordinary-looking squashes are hugely popular in Japan and it is arguably as important a part of Japanese food culture as that of North America and southern Europe. Regular Japanese cultivars in my garden are *Uchiki Kuri* (*kuri* means 'chestnut' in Japanese) and *Kabocha*.

And They Grew – and They Grew

The fascination with growing giant specimens of *C. maxima* has remained something of an obsession for many during the last 500 years. Yet again, lazy journalists and seed sellers frequently refer to 'giant pumpkins' or 'the world's largest pumpkin'. According to the *Guinness Book of Records*, the biggest specimen, which is a record that has been held since 2016, was grown by Mathias Willemijns of Ludwigsburg, Germany. Weighing in at 1,190.49 kg (2,625lb), it is the size of a small car.[9] One thing this inedible behemoth isn't, however, is a pumpkin.

In concluding this chapter, I feel compelled to return to the vexed theme: when is a pumpkin really a pumpkin? What has been agreed by most commentators since the word 'squash' came into common parlance, is that all pumpkin are types of squash but only one type of squash can be called a pumpkin. The botanist W.F Giles devoted five pages to the question of naming and identifying different species of cucurbits in the *Journal of the Royal Horticultural Society* in May 1943.[10] In the USA popular classifications have never agreed with the botanical ones and, as we have seen, pumpkin is a word invented by the British as a general description of all types of squash – itself the anglicisation of a Native American word that described all types of cucurbit under cultivation in North America.

I take my lead from David Landreth who, along with his brother Cuthbert, founded a seed company named after them in 1784. One of the first seed companies in the US, it became synonymous with promoting heritage and heirloom American vegetable varieties. David perfectly described the difference between a squash and a pumpkin thus: 'For practical purposes the farmer's test is the best, and that is, when he can stick his thumb-nail into the rind after reaching full development, it is a Pumpkin, when he cannot, it is a Squash, as its rind becomes as hard as wood'. There is only one type of squash that goes by the botanical name pumpkin, and that is the one identified by Harry S. Paris. This is a soft-skinned summer squash belonging to the species *C. pepo*. The most famous so-called pumpkin, Jack O' Lantern, as Sauer pointed out is, a variety of hard-skinned *C. pepo*. However, its skin is not so hard that it cannot be called a pumpkin – which it is.

Is this really a lot of fuss about nothing? In the great scheme of things, probably. But, for me, how we name our crops is key to our relationship with them. What they are called matters not just to me but to everyone who feels a strong connection with what they grow. Your pumpkin may be my squash, so let's not fight over the name and just celebrate their unique place in the story of their domestication, eating them, carving faces in them, encouraging them through diligent husbandry to achieve a vast size and, most of all, enjoying their flesh, their seeds and their flowers.

And Finally
– Seeds of Hope

Make hunger thy sauce, as a medicine for health
Thomas Tusser (1524–1580) – *Five
Hundred Points of Good Husbandry* (1573)

W hen I started to grow vegetables on a few acres of the family farm in Devon in the late 1970s, no one was interested in buying red Brussels sprouts or yellow courgettes. The fact that they were 'organic' was met with some suspicion. After a couple of hard winters, the pleasure of picking sprouts late into a freezing evening to flog them next day in the local market for a pound or two wore off. Better to grow for love than for a living. I saw myself then as part of a hippy revolution to turn the world green; I was one of a tiny minority of growers. In those days, people who were attempting to grow food in a sustainable way were generally ignored; the public appeared completely uninterested. It was to take another couple of decades for attitudes to really start to change.

Rolling Back the Years

Vegetables have been lost or become extinct for a variety of reasons, as we have seen. And it is not just due to a loss of habitat:

267

in many instances they have simply fallen out of culinary favour or been the victims of commercial abandonment. We are today paying the price for a binary approach to feeding ourselves: increase yield at all costs and as cheaply as possible without any regard for the environment. The triumph of quantity over quality has resulted in us eating food of poorer nutritional value than that enjoyed by our parents and grandparents. Despite many modern cultivars of popular vegetables being more vigorous, having heavier yields and greater disease resistance, my taste buds tell me they usually have less flavour. We have become used to buying and eating food that is unripe and tasteless, despite what marketeers might say to the contrary.

Today, however, quality, nutritional value and provenance are among the key factors driving a change in public opinion. We are now re-learning traditional ways to produce our food, which were developed over millennia by farmers across the globe. Ignored and denigrated in the last century, a sustainable, holistic and inclusive model for growing our food is no longer a choice; it is a necessity. For the first time that I can remember, traditional varieties are gaining in popularity. Consumers are experiencing the taste of sustainably grown food and demanding more. There is compelling evidence that the world values its food culture and seeks to revive and support it – a prime example being India, where I have witnessed the knowledge and traditions of peasant farmers at the forefront of this change.

A quarter of a century ago, research and development on how to feed the planet with a burgeoning population in the face of climatic challenges was focused on technological solutions. Now, an ever-growing number of us are recognising and respecting the real value of biodiversity and sustainable and traditional forms of agriculture.

This recognition applies equally to the unmatched knowledge of peasant farmers and the rights they should be entitled to in respect of their crops. The expertise of peasant farming is becoming ever more respected, and I sit at the feet of these people who are passing on millennia of understanding and practice. They are a fundamental part of the solution to feeding the world.

A Sustainable and Affordable Future

The real danger I see today is an increasing divergence between those who can afford to eat organic and sustainably produced healthy food and those who can only afford the very cheapest and poorest quality, mostly processed, unhealthy food. One of the main challenges facing society is to ensure that everyone, regardless of their income, has access to affordable and nutritious food. A new generation of very determined growers has embraced regenerative horticulture and taken on the task of addressing this challenge. They give me hope because growing fruits and vegetables organically requires fewer inputs and therefore lower overheads. Not only that, but market gardens and small-scale horticulture enterprises growing a diversity of crops can generate a net income of £20,000 (approx. $26,000) an acre – ten times as much as large-scale intensive horticulture. Veg-box schemes where local consumers receive a weekly supply of what the grower has harvested, farmers markets and farm-gate sales, where people engage with farmers and understand their work, transform public perception and are invariably better value than supermarket offerings. With scale and shorter supply chains will come further savings. The hegemony of the supermarket is being challenged and every day more people are finding locally grown food affordable.

269

The new generation of growers, for whom I have such admiration, are committed to producing food sustainably, celebrating the diversity and richness of crop varieties, many of which have been part of our diet for centuries. We may well face an existential threat to our food supply due to the iniquities of modern food production, but the solutions to this crisis are in our hands, and in writing this book I hope to have shared my excitement for re-establishing and strengthening our lost relationship with what we grow. By taking a holistic approach to resolving these challenges and exploiting the best that scientific research, human ingenuity and a sensitivity to the living world has to offer, we really can provide our species with the diverse, healthy and sustainable diet that it needs.

The tide is turning against a policy of monoculture. Exciting research now shows that nations which grow a greater diversity of crops are able to mitigate against many of the effects of global warming. The greater the diversity, the greater the stability of harvests in the face of weather extremes, be it drought or flood.[1] The 'ground up' revolution in sustainable agriculture across the world is gaining more and more traction. The present unsustainable and inequitable models of producing food, where farmers receive the least reward for their labours, and most of the profits are in the hands of distributors and retailers, are also being challenged. Farmers are collaborating more effectively to take back control of the means of production, distribution and sale, so they can be partners in building sustainable and profitable businesses, growing the food that keeps us healthy. Today, this planet is more than capable of feeding itself. The discredited paradigm that has underpinned the approach to food production we have seen since World War II must continue to be strictly regulated, challenged and reformed.

And let's not forget amateur gardeners, whether they are harvesting herbs from a pot on the windowsill or eating produce from their gardens and allotments, who are also part of the solution. There are more than a million acres of gardens in the U.K., which represents 8 per cent of all land used for growing crops: about the same percentage as land under commercial organic cultivation. These can be the most biodiverse spaces for plants, wildlife and food production, growing and saving seeds of local and culturally important crops.

The history of plant breeding is littered with stories of the abuse and exploitation of the genetic resources of indigenous peoples. There is no more invidious example of this than the appropriation of unique Folk Varieties that have formed the basis for the breeding of modern cultivars which have then been claimed as the intellectual property of the seed producer. The farmers who have grown the parent for generations receive no acknowledgement or financial reward. Often these modern cultivars, many genetically modified, are sold back to the native farmer at great cost and with no possibility for them to save the seed, as this would be a breach of the seed producer's intellectual property rights. This perverse situation is at last being challenged. Seed banks, institutions that maintain collections of edible crop species which can be returned to growers to bulk up in the event of total loss; libraries that maintain living collections of heritage and heirloom varieties, which are shared or loaned to growers to maintain them; and plant breeders developing new cultivars have signed up to a number of protocols with organisations, including the UN and the International Seed Federation, to respect the creation and conservation of genetic material for the benefit of all. Now, at last, the importance of conserving and strengthening the genetic diversity of crops is recognised as being crucial to combating climate change and feeding the world. Traditional varieties are becoming more highly valued because they are a critical part of the solution to food security as well as being a valuable genetic resource for plant breeders.

Today, transferring genetic material, which includes seeds, between countries is highly regulated. When I began writing this book, moving vegetable seeds across borders within the EU happened freely. Now it is only possible with much paperwork and certification. This is also the case globally. Businesses can no longer collect native varieties of heirlooms and indigenous varieties and bring them home to grow for commercial exploitation.

Although vegetable seeds are not likely carriers of pathogens and bugs that can devastate native species, I am very careful about what I bring home. Certified commercial vegetable seed needs to be cleared through customs but is not subject to the protocols around FV.* Bringing seeds across

---

* The first global multilateral treaty for the conservation and sustainable use of genetic plant material, The Convention on Biological Diversity (CBD), was drafted by the United Nations in 1992. Article 15 of the CBD specifies that nations have sovereignty over their genetic resources but that there should be access and benefit-sharing by mutual agreement with prior informed consent. The International Treaty on Plant Genetic Resources for Food and Agriculture (ITP-GRFA) was adopted by the Food and Agricultural Organisation (FAO) of the UN and came into operation in 2004. The treaty stops recipients of genetic resources from claiming intellectual property rights over them and provides a legal framework in which donor and recipient can operate, known as the Standard Material Transfer Agreement (SMTA).

borders, although highly regulated, is not always policed and I would be lying if I said that this has not sometimes been to my benefit. I follow my own strict protocol regarding heritage and heirloom seeds. I always return seeds whenever possible to the person who gave them to me. I always seek the permission of the person who gave me the seeds before sharing them with others. This is an important demonstration of the circularity of our food. I never share seeds that could be commercially exploited because those seeds are only ever loaned to me. I don't own them.

My next seed-hunting trips will focus on encouraging growers to maintain local varieties and learning from the new generation of plant breeders who are finding ways to ensure the survival of rare and endangered varieties in their country of origin. Of course, if I come across highly endangered varieties or am offered seed of some wonderful and delicious local vegetable to try, I'll probably find a corner of my suitcase to fit in a few, but whether that would be something I declared on arrival back in the U.K. I could not possibly comment upon.

....

We may have lost some 90 per cent of all varieties of fruit and vegetables in the last century but plant breeders continue to develop hundreds of new cultivars of pretty much every type of crop, many of which can help global food security. But, for me, the real heroes of our future food supply are the growers, farmers and seed libraries who are recovering, restoring and championing local varieties around the world. Despite the catastrophic loss of genetic diversity, advances in plant genetics and a focus on working with the wild relatives of edible plants to create new cultivars is very exciting. The wonders I grow in my garden and delight in eating are only with me because of a long string of actions taken by growers, breeders and communities to conserve the food that is at the core of their sense of self. We are now, I hope, on a journey back to a more meaningful relationship with our soil, our seeds and our produce. Long may it continue.

Acknowledgements

I have my late mother, June, to thank for instilling in me a desire to grow stuff. She was an opinionated and sometimes careless gardener but her connection with the soil was as an umbilical cord and through it flowed a love of cultivation that infected not only me but my siblings too. I also have to thank an anthroposophical education with an enlightened approach to detention – time spent in the school's walled garden – for instilling in me a questioning of horticultural orthodoxy and a desire to embrace the widest selection of ideas and approaches to growing food. Despite a lifetime in the garden and half a lifetime collecting and saving seeds, the genesis for this book came only in 2013. I met up with an old pal who thought I might like to step out from behind the camera and travel the world to uncover stories behind how the vegetables that I love to grow and save seed from found their way into my garden. It was another four years before I finally accepted that broadcasters did not share our vision. But that did not deter me from spending the following four years doing the thing I had always wanted to do: share my curiosity about the how's and whys of our relationship with what we grow and eat through writing.

The last few years have been a hugely enjoyable journey, a giant learning curve, an unforgettable adventure, thanks in every way to friends, family and colleagues. I am indebted to Paul Manias who believed others might be interested to know about my world as the seed detective and started me on this journey; to Mark Gould for his creativity and optimism and continued belief that there is a TV series in this; to my good friend the historian Will Davies for sharing his way into research and building early drafts; to Neil Fearis for reading every word and reminding me, through his uninhibited use of a red pencil, of what I was meant to have learned in English classes when we were at school together; for encouragement, advice and guidance from Broo Doherty; to Thomas Stäedler for sharing his knowledge and expertise of wild tomatoes; to Svend Erik Nielsen for sharing his curiosity on the provenance of peas; to my colleague Bruce Paterson at Garden Organic; to my dear friend the writer Andrew Taylor, my mentor and sage on all matters relating to the world of publishing, who helped me believe I could write this; to Pauline Lee for her invaluable early research; my son Jesse and friends Roger Moore, David Hutt and Colin Luke who took the time to read various drafts and sketches and whose advice I appreciated even if I frequently ignored it; to Phillipa and Bex Higson for their detective quote-hunting; to Muna Reyal, who as all good editors should, helped to make the work better, along with the inspiring team at Chelsea Green U.K.: Matt Haslum, Alex Stewart, Rosie Baldwin and Harshita Lalwani; not forgetting my copy editor Caroline West whose forensic interrogation of the manuscript reminded me of my English essay homework returns but more helpful; to my agent Sonia Land, who, when we weren't discussing our horticultural failures and successes kept me focused in the dark times; and to my other son Jake, a professor of plant ecology who patiently and diligently shared his botanical knowledge with me and did his best to keep me on the scientific straight and narrow.

Acknowledgements

My thanks to Alan Kapuler for sharing his adventures with breeding peas and to the Cardiff Naturalists' Society for permission to use extracts from their magazine. I am especially grateful to the 'jewel in the crown' of the charity Garden Organic – the Heritage Seed Library – Catrina Fenton and those of her amazing team who have supported, supplied and tolerated my endless requests not only for seeds but their knowledge; as well as Rachel Crow and Lucy Shepherd.

To those on whose wisdom I have depended, if I have got things wrong, I apologise. All errors of historical and botanical matters are entirely down to me.

Above all, I want to thank my beloved Julia for believing in me, for her daily encouragement, for reading the manuscript in its various iterations, and for patiently making me think about what I wanted to say and how to write it. It is because of her invaluable, honest and uncompromising comments and observations that this book exists at all.

Glossary

Allele: The name given to a variant form of a gene located at the same position – the genetic locus – on a chromosome. Vegetables that are described botanically as *diploid* have two alleles at each locus with one allele inherited from each parent. Each pair of alleles represents the *genotype* of a specific gene, being described as *homozygous* if the alleles at a particular locus are identical and *heterozygous* if different. Genes also contribute to an organism's outward appearance known as its *phenotype*.

Annual: A vegetable that grows from seed and flowers in the same year.

Archaeobotany: A subdiscipline of archaeology in which aspects of the human past are interpreted through the examination of preserved plant remains.

Biennial: A vegetable that takes two years to grow, flower, set seed and die.

Centre of Origin: Describes a geographical area where a plant species, either domesticated or wild, first developed its distinctive properties.

Clone: An organism that is genetically identical to the individual from which it was propagated. For example, garlic grown from a clove is a clone of the original bulb.

Cross-pollination: The result of one variety (or individual) of a species pollinating a different variety (or individual). The genetic material of the two different varieties combines, resulting in a new variety with characteristics inherited from both its parents.

Cultivar: A cultivated variety that has been bred either vegetatively by grafting or is the result of an accidental or deliberate mutation or hybridisation. All F1 hybrid vegetables are cultivars which do not generally breed true from seed. Its continued existence is wholly dependent upon human intercession, just as 'cultivated' land will revert to a wild state without continuous human action.

DNA fingerprinting or **DNA barcoding:** The process of determining an individual organism's identity based on its DNA. A short section – sequence – of DNA from a gene or part of a genome is compared with a similar sequence held in a genetic reference library. This enables plant scientists to identify varieties and also determine if a variety is unique and distinct or synonymous with another, differently classified variety of the same species.

Domestication: The outcome of a selection process that leads to plants adapted to cultivation and utilisation by humans.

Domestication syndrome: Typical traits of domesticated plants that differentiate them from their wild ancestors.

Fertile Crescent: The region of South-West Asia, comprising the valleys of the Tigris, Euphrates and Jordan rivers and their adjacent hilly flanks, which is one of the global origins of settled agriculture.

Genome: The complete set of an organism's DNA, including all its genes.

Gene editing: Also known as genome editing, this describes a process in which DNA can be inserted, modified, deleted or used to replace genes at specific locations within the genome of an organism. It is unlike other forms of genetic engineering which insert foreign genes into the organism's genome.

Gene flow: The transfer of alleles of a gene from one population to another, for example by cross-hybridisation (interbreeding).

Gene mapping: The process of determining the location of genes on the chromosomes of a species.

Genetically Modified Food (GMF): Refers to plants that have had changes made to their DNA using genetic engineering techniques in order to introduce specific traits, including resistance to specific herbicides and pesticides.

Heirloom: Describes a variety of vegetable that has usually arisen accidentally and been grown for several generations with a provenance that is either specific to a place or a people or both. The term is also used loosely to describe old-fashioned or traditional varieties that have been commercially revived.

Heritage: Describes a fruit or vegetable that has a cultural significance because of where it is cultivated and how it is used. The word is also used to describe some ex-commercial varieties that are conserved within gene banks and seed libraries. A number of heritage varieties are being revived commercially.

Heterogeneous: When a variety or species displays genetically determined variability in traits or attributes (such as flowering period, disease resistance, seed size), which might be of interest to plant breeders.

Hybrid: The outcome of deliberate crossbreeding of plants from different varieties (or even species) in order to create a cultivar with specific traits, such as F1 hybrids. Natural hybridisation occurs when individuals from two distinct populations or groups of the same species, which are distinguishable on the basis of one or more heritable characteristics, interbreed.

Inbreeder: A species that is *self-fertile* and hence able to pollinate itself, resulting in inbreeding. Known also as selfers or self-pollinators, the most common *self-compatible* or *interfertile* vegetables of this type include peas, common beans and most tomatoes. Flowers of many inbreeders can also be cross-fertilised by the action of pollinators like bees.

Indigenous: Describes a species that is *native* to (has evolved naturally in) a particular geographic area.

Introgression: The process by which genes of one species (or variety) enter the genome of another species (or variety) through repeated rounds of interbreeding.

Landrace: A locally distinct, and potentially adapted, population of a crop, often associated with traditional farming systems.

Mesoamerica: A region of Central America that extends from central Mexico southwards and includes Costa Rica, Nicaragua, Honduras, Guatemala and Belize. It is one of the world's most important centres of diversity for domesticated crops.

Monophyletic: Refers to a group of organisms or DNA sequences that are derived from a single ancestral organism or DNA sequence.

Morphology: The structure, form and physical features of a plant.

Mutation: Describes a change in the DNA sequence of an organism, which arises naturally due to errors in DNA replication or from damage to DNA. Mutations can also be deliberately introduced by plant breeders using chemicals or radiation to create new cultivars.

Neolithic: The period of human cultural development characterised by the adoption of a subsistence culture based on agriculture.

Open-pollinated: This term describes plants that are pollinated by wind or animals. Such varieties are said to 'breed true', meaning the progeny are identical to the parent.

Outcrossing: Also known as ***outbreeding***, this is the result of fertilisation between different individuals, including distant relatives (compare with *Hybridisation*). The most important group of vegetable that is the result of outcrossing between different species are members of the brassica family.

Perennial: A vegetable that lives for more than two years.

Phenotype: The observable characteristics or traits of a species, including its physical form and structure (***morphology***).

Polymorphism: The occurrence of two or more different forms, also known as alternative phenotypes, in the population of a species.

Variety: Used here to mean an alternative form of a plant that arises from accidental crossing between other varieties of the same species. In contrast to cultivars, varieties are not normally bred deliberately and usually breed true without human intervention; in other words, the offspring are identical to the parent and share its unique characteristics and traits.

Notes

Introduction

1. Vavilov's seminal work where he identifies 'Centres of Origin': N. I. Vavilov, *Origin and Geography of Cultivated Plants*, trans. Doris Löve (Cambridge, U.K.: Cambridge University Press, 1992).
2. An interesting article on the question of just how many Centres of Diversity exist: K. Kris Hirst, 'The Eight Founder Crops and the Origins of Agriculture', *ThoughtCo*, 31 August 2018, https://www.thoughtco.com/founder-crops-origins-of-agriculture-171203.
3. Amanda J. Landon, 'The "How" of the Three Sisters: The Origins of Agriculture in Mesoamerica and the Human Niche', *Nebraska Anthropologist* 23 (2008): 40; Paul Gepts, 'Crop Domestication as a Long-Term Selection Experiment', *Plant Breeding Reviews* 24, no.2 (2004): 1–44.

Part One – Visitors from the East

1. *Meet the Romans with Mary Beard*, Episode 1, 'All Roads Lead to Rome', Lion TV, first broadcast on BBC Television 2012.

The Tale of Four Peas or Four Tall Stories

1. Tony Winch, *Growing Food: A Guide to Food Production* (Herefordshire, U.K.: Clouds Books, 2014): 160–61.
2. *Norman F. Weeden, 'Domestication of Pea (Pisum sativum L.): The Case of the Abyssinian Pea', Frontiers in Plant Science* 9, (2018): 515, https://doi.org/10.3389/fpls.2018.00515.including groups a-1, a-2, b, c, and d as identified by Kwon et al. (2012).

283

3. More on the Carlin pea: 'Carlin Peas, a Northern Tradition', *Heritage and History*, 5 April 2010, https://www.heritageandhistory.com/contents 1a/2010/04/carlin-peas-a-northern-tradition.
4. Quackwriter, 'Mr Grimstone and the Revitalised Mummy Pea', *The Quack Doctor*, 14 January 2014, http://thequackdoctor.com/index.php /mr-grimstone-and-the-revitalised-mummy-pea.
5. Jonathan D. Sauer, *Historical Geography of Crop Plants: A select roster* (Boca Raton, FA: CRC Press, 1993): 67–69.

A Broad Bean Far from Home
1. Dorian Q. Fuller, George Willcox and Robin G. Allaby, 'Early Agricultural Pathways: Moving Outside the "Core Area" Hypothesis in Southwest Asia', *Journal of Experimental Botany* 63, no. 2 (January 2012): 617–33, https://doi.org/10.1093/jxb/err307.
2. Valentina Caracuta et al., '14,000-Year-Old Seeds Indicate the Levantine Origin of the Lost Progenitor of Faba Bean', *Scientific Reports* 6 (November 2016): 37399, https://doi.org/10.1038/srep37399.
3. Oleg Kosterin, 'The "Lost Ancestor" of the Broad Bean (*Vicia faba* L.) and the Origin of Plant Cultivation in the Near East' [in Russian], *Vavilov Journal of Genetics and Breeding* 18, no. 4 (2014): 831–40, https://doi.org /10.18699/VJ15.118.
4. Fuller, Willcox and Allaby, 'Early Agricultural Pathways': 617–33.
5. More on the etymology behind naming plants: 'Fava (n.)', *Etymology Online*, last accessed 14 March 2022, https://www.etymonline.com/word/fava.
6. More about the goddess Carna: 'Carna', *The Obscure Goddess Online Dictionary*, last accessed 14 March 2022, http://www.thaliatook.com/OGOD/carna.php.

Orange is Not the Only Colour
1. O. Banga, 'Origin and Distribution of the Western Cultivated Carrot', *Wageningen*, no. 222 (1964): 357–70.
2. Nikolai Ivanovich Vavilov, 'Studies on the Origin of Cultivated Plants', *Bulletin of Applied Botany and Plant Breeding*, no.1 (1926).
3. John Stolarczyk and Jules Janick, 'Carrot: History and Iconography', *Chronica Horticulturae* 51, no. 2 (2011): 16.
4. Kassia St Clair, *The Secret Lives of Colour* (London: John Murray, 2016): 88.
5. Banga, 'Origin and Distribution': 357–70.

6. P.R. Ellis et al., 'Exploitation of the Resistance to Carrot Fly in the Wild Carrot Species *Daucus capillifolius*', *Annals of Applied Biology* 122, no. 1 (February 1993): 79–91.

In Search of a Welsh Leek

1. Christopher D. Preston, David A. Pearman, and Allan R. Hall, 'Archaeophytes in Britain', *Botanical Journal of the Linnean Society* 145, no. 3 (July 2004): 257–94.

2. Yann Lovelock, *The Vegetable Book: An Unnatural History* (London, U.K.: George Allen & Unwin, 1972): 158.

3. Eleanor Vachell, 'The Leek: The National Emblem of Wales', *Transactions of the Cardiff Naturalists' Society* 3, (1993): 26.

4. James L. Brewster, *Onions and Other Vegetable Alliums* (Wallingford, U.K.: CABI, 2008).

5. Lovelock, *The Vegetable Book*, 158.

6. Lovelock, *The Vegetable Book*, 158.

7. An article giving an account of the 25th anniversary of this discovery: Fred Searle, 'Breeders Celebrate 25th Anniversary of First Hybrid Leek', *Fresh Produce*, 5 October 2018, http://www.fruitnet.com/fpj/article/176813 /breeders-celebrate-25th-anniversary-of-first-hybrid-leek.

Of Caulis, Krambē and Braske

1. Lorenzo Maggioni, 'Domestication of *Brassica oleracea* L.', (PhD diss., Swedish University of Agricultural Sciences, 2015), 74, https://pub.epsilon .slu.se/12424/1/maggioni_l_150720.pdf.

2. Rebecca Rupp, *How Carrots Won the Trojan War: Curious (but True) Stories of Common Vegetables* (North Adams, MA: Storey, 2011): 67.

3. Johannes Helm, 'Morphologisch-taxonomische Gliederung der Kultursippen von *Brassica oleracea* L.', *Die Kulturpflanze* 11 (1963): 92–210.

4. J.T.B. Syme, 'Brassica Oleracea', *English Botany* 1, no. 3 (1863): 130–33.

5. N.D. Mitchell, 'The Status of *Brassica oleracea* L. Subsp. *Oleracea* (Wild Cabbage) in the British Isles', *Watsonia* 11 (1976): 97–103.

6. Edward Lewis Sturtevant, *Sturtevant's Edible Plants of the World*, ed. U.P. Hedrick (Geneva, NY: New York Agricultural Research Station, 1919), https://www.swsbm.com/Ephemera/Sturtevants_Edible_Plants.pdf.

7. Rupp, *How Carrots Won the Trojan War*. 71.

8. Lee B. Smith and Graham J. King. 'The Distribution of BoCAL-a Alleles in *Brassica oleracea* Is Consistent with a Genetic Model for Curd Development and Domestication of the Cauliflower', *Molecular Breeding* 6 (2000): 603–13, https://doi.org/10.1023/A:1011370525688.
9. Lovelock, *The Vegetable Book*, 72.

An Aspiring Spear

1. Alfred W. Kidner, *Asparagus* (London, U.K.: Faber, 1959): 19.
2. Sutton & Sons, *The Culture of Vegetables and Flowers: From Seeds and Roots* (Reading, U.K.: Sutton & Sons, 1884): 7.
3. Kidner, *Asparagus*, 21.
4. This story and much more about asparagus in America: Ruth Lyon, 'Stalking Diederick's Asparagus', *Asparagus Lover*, last accessed 14 March 2022, https://www.asparagus-lover.com/diedericks-asparagus .html.
5. Kidner, *Asparagus*, 25.
6. More can be read on this subject here: Rachel Dring, 'Asparagus: Draining Peru Dry', *Sustainable Food Trust*, 17 April, 2014, https://sustainablefood trust.org/articles/asparagus-draining-dry.
7. Peter G. Falloon, 'The Need for Asparagus Breeding in New Zealand', *New Zealand Journal of Experimental Agriculture* 10, no. 1 (January 1982): 101–9, https://doi.org/10.1080/03015521.1982.10427851.
8. Another great source of information on all things asparagus: http://www .britishasparagusfestival.co.uk.

For the Love of a Leaf

1. I.M. de Vries, 'Origin and Domestication of *Lactuca sativa* L.', *Genetic Resources and Crop Evolution* 44, no. 2 (April 1997): 165–74, https://doi .org/10.1023/A:1008611200727.
2. Ludwig Keimer, *Die Gartenpflanzen im alten Agypten. Band I* [Garden plants in ancient Egypt: Volume 1] (Hamburg, Germany: Hoffman und Campe Verlag, 1924), 187.
3. Lovelock, *The Vegetable Book*, 125.
4. Jack R. Harlan, 'Lettuce and the Sycomore: Sex and Romance in Ancient Egypt', *Economic Botany* 40 (January 1986): 4–15, https://doi.org /10.1007/BF02858936.

5. P.H. Oswald, 'Historical Records of *Lactuca serriola* L. and *L. virosa* L. in Britain with Special Reference to Cambridgeshire', *Watsonia* 23 (2000): 149–59.

6. Robert James Griesbach, *150 Years of Research at the United States Department of Agriculture: Plant Introduction and Breeding* (Beltsville, MD: USDA, 2013), https://www.ars.usda.gov/ARSUserFiles/oc/np/150YearsofResearchatUSDA/150YearsofResearchatUSDA.pdf.

7. More on this subject: 'Lettuce "Great Lakes"', *Seedaholic*, last accessed 14 March 2022, http://www.seedaholic.com/lettuce-great-lakes.html.

8. H.A. Jones, 'Pollination and Life History Studies of Lettuce (*Lactuca sativa* L.)', *Hilgardia* 2, no. 13 (April 1927): 425–79, https://doi.org/ 10.3733/hilg.v02n13p425.

9. Lovelock, *The Vegetable Book*, 90.

Thank Goodness for Garlic

1. Ahmed Nasser Al-Bakri et al., 'The State of Plant Genetic Resources for Food and Agriculture in Oman', (Directorate General of Agriculture and Livestock Research, 2008): 13, https://doi.org/10.13140/RG.2.2.27934.33609.

2. T. Etoh and P.W. Simon, 'Diversity, Fertility and Seed Production of Garlic Allium' in *Allium Crop Science: Recent Advances*, ed. H.D. Rabinowitch and L. Currah (New York: CABI, 2002): 101, https://doi.org/10.1079/9780851995106.0000.

3. K. Kris Hirst, 'Garlic Domestication – Where Did It Come from and When?', *ThoughtCo.*, updated 28 July 2019, https://www.thoughtco.com/garlic-domestication-where-and-when-169374.

4. Larry D. Lawson 'Garlic: A Review of Its Medicinal Effects and Indicated Active Compounds', *Phytomedicines of Europe. American Chemical Society* (1998): 176–209.

5. R.S. Rivlin, 'Historical Perspective on the Use of Garlic', *Journal of Nutrition* 131, no. 3 (2001): 952S.

6. More on this subject can be found at: http://www.krishna.com/why-no-garlic-or-onions.

7. Philipp W. Simon, 'The Origins and Distribution of Garlic: How Many Garlics Are There?' USDA Agricultural Research Service, updated 3 March 2020, https://www.ars.usda.gov/midwest-area/madison-wi/vegetable-crops-research/docs/simon-garlic-origins.

8. Simon, 'The Origins and Distribution of Garlic'.
9. Alessandro Bozzini, 'Discovery of an Italian Fertile Tetraploid Line of Garlic', *Economic Botany* 45, no. 3 (1991): 436–38.

Part Two – Arrivals from the West

1. Richard S. MacNeish, Antoinette Nelken-Terner and Irmgard Weitlaner Johnson, *The Prehistory of the Tehuacán Valley* (Austin, TX: University of Texas Press, 1967): 220–23; James F. Hancock, *Plant Evolution and the Origin of Crop Science*, 2nd ed. (Wallingford, U.K.: CABI Publishing, 2004): 157.
2. John M. Kingsbury, 'Christopher Columbus as a Botanist', *Arnoldia* 52, no. 2 (1992), http://www.arnoldia.arboretum.harvard.edu/pdf/articles /1992-52-2-christopher-columbus-as-a-botanist.pdf.

More Than Just A Fruit

1. Nicolas Ranc et al., 'A Clarified Position for *Solanum lycopersicum* Var. *cerasiforme* in the Evolutionary History of Tomatoes (Solanaceae)', *BMC Plant Biology* 8, no. 1 (2008): 130, http://doi.org/10.1186/1471-2229-8-130.
2. Jose Blanca et al., 'Variation Revealed by SNP Genotyping and Morphology Provides Insight into the Origin of the Tomato', *PLOS ONE* 7, no. 10 (2012): e48198, https://doi.org/10.1371/journal.pone.0048198.
3. Sauer, *Historical Geography of Crop Plants*: 156.
4. J.A. Jenkins, 'The Origin of the Cultivated Tomato', *Economic Botany* 2, no. 4 (1948): 379–92.
5. Sauer, *Historical Geography of Crop Plants*, 157.
6. The first recipe for 'katsup' appeared in Eliza Smith, *The Compleat Housewife, or, Accomplished Gentlewoman's Companion* (London: Pemberton, 1727), https://archive.org/details/2711361R.nlm.nih.gov/mode/2up.
7. David Gentilcore, *Pomodoro!: A History of the Tomato in Italy* (New York: Columbia University Press, 2010): 11.
8. A very good description of the different flower and leaf types of tomato: 'Exploring Tomato Flower Structure', Seed Savers Exchange (blog), 18 April 2016, https://blog.seedsavers.org/blog/exploring-tomato-flower -structure.
9. 'Fine Tomatoes', *American Agriculturist* (October 1869), 362. A detailed description of the Trophy tomato can be found here: https://www .biodiversitylibrary.org/item/245834#page/1/mode/1up

10. Liberty Hyde Bailey, *The Survival of the Unlike: A Collection of Evolution Essays Suggested by the Study of Domestic Plants* (USA: Palala Press, 1897): 1858–954.

11. 'Global Tomato Seed Market Analysis & Outlook 2019–2024 – The Intensification of Hybrid Seed Usage across Emerging Economics Is Driving Growth', *Business Wire*, 3 April 2019, https://www.businesswire.com/news/home/20190403005336/en/Global-Tomato-Seed-Market-Analysis-Outlook-2019-2024---The-Intensification-of-Hybrid-Seed-Usage-across-Emerging-Economics-is-Driving-Growth---Research AndMarkets.com.

12. R.A. Jones and S.J. Scott, 'Improvement of Tomato Flavor by Genetically Increasing Sugar and Acid Contents', *Euphytica* 32, no. 3 (1983): 845–55.

A Very Uncommon Bean

1. Everything you might be curious to know about this bean and how the local community has built a sustainable business model with it: Conzorzio per la Tutela del Fagilo di Lamon della Ballata Bellunese IGP [Consortium for the Protection of the Lamon Bean of the Belluno Valley], last updated 2013, http://www.fagiolodilamon.it.

2. This poem appears in: Ken Albala, *Beans: A History* (London: Bloomsbury, 2007): 120.

3. Lawrence Kaplan, 'What Is the Origin of the Common Bean?', *Economic Botany* 35, no. 2 (April 1981): 240–54, https://doi.org/10.1007/BF02858692.

4. Paul Gepts, 'Origin and Evolution of Common Bean: Past Events and Recent Trends', *Hortscience* 33, no. 7, (December 1998): 1124–30, https://doi.org/10.21273/HORTSCI.33.7.1124; Paul Gepts, '*Phaseolus vulgaris* (Beans)', *Encyclopedia of Genetics* (December 2001): 1444–45, https://doi.org/10.1006/rwgn.2001.1749.

5. Kaplan, 'What Is the Origin of the Common Bean?': 240–54.

6. Albala, *Beans*: 118.

7. The Herbal or Generall Historie of Plants by John Gerarde can be found online T https://archive.org/details/mobot31753000817749/mode/1up?ref=ol&view=theater. The Second Booke of the Historie of Plants Chapter 490, 1038-1042

8. Albala, *Beans*: 125.

The Colour of Corn

1. Daniela Soleri and David A. Cleveland, 'Hopi Crop Diversity and Change', *Journal of Ethnobiology* 13, no. 2 (1993): 209.
2. Daniela Soleri and Steven E. Smith, 'Morphological and Phenological Comparisons of Two Hopi Maize Varieties Conserved in situ and ex situ', *Economic Botany* 49 (January 1995): 56–77, https://doi.org/10.1007/BF02862278.
3. William L. Brown, E.G. Anderson and Roy Tuchawena Jr, 'Observations on Three Varieties of Hopi Maize', *American Journal of Botany* 39, no. 8 (October 1952): 597–609, https://doi.org/10.2307/2438708.
4. Feng Tian, Natalie M. Stevens and Edward S. Buckler IV, 'Tracking Footprints of Maize Domestication and Evidence for a Massive Sweep on Chromosome 10', *PNAS* 106, supplement 1 (16 June 2009): 9875, https://doi.org/10.1073/pnas.0901122106.
5. P.C. Mangelsdorf and R.G. Reeves, 'The Origin of Maize', *PNAS* 24, no. 8 (15 August 1938): 304–6, https://doi.org/10.1073/pnas.24.8.303.
6. G.W. Beadle, 'Teosinte and the Origin of Maize', *Journal of Heredity* 30, no. 6 (1939): 245–47, https://doi.org/10.1093/oxfordjournals.jhered.a104728.
7. David A. Cleveland, Daniela Soleri and Steven E. Smith, 'Do Folk Crop Varieties Have a Role in Sustainable Agriculture? Incorporating Folk Varieties into the Development of Locally Based Agriculture may be the Best Approach', *BioSience* 44, no. 11 (December 1994): 743, https://doi.org/10.2307/1312583.
8. Jock R. Anderson, Peter B.R. Hazell, and Lloyd T. Evans, 'Variability of Cereal Yields: Sources of Change and Implications for Agricultural Research and Policy', *Food Policy* 12, no. 3 (August 1987): 199–212, https://doi.org/10.1016/0306-9192(77)90021-5.
9. A good starting point to explore more about the challenges and complexities of maize as a monocrop and biofuel: Jonathan Foley, 'It's Time to Rethink America's Corn System', *Scientific American*, 5 March 2013, https://www.scientificamerican.com/article/time-to-rethink-corn.
10. Tian, Stevens and Buckler, 'Tracking Footprints of Maize Domestication': 9984.

The Tale of Two Classy Beans

1. Winch, *Growing Food*: 174–77.

2. Sara Roahen, 'Christmas Lima Beans', Slow Food USA, 20 February 2018, https:// slowfoodusa.org/christmas-lima-beans.
3. Sauer, *Historical Geography of Crop Plants*, 77–80.
4. Sauer, *Historical Geography of Crop Plants*, 77–80.
5. Kenneth F. Kipple, *A Moveable Feast: Ten Millennia of Food Globalization* (New York: Cambridge University Press, 2007): 115.

Some Like 'em Hot

1. B. Pickersgill, C.B. Heiser Jr and J. McNeill, 'Numerical Taxonomic Studies on Variation and Domestication in Some Species of Capsium', in eds. J.G. Hawkes, R.N. Lester and A.D. Skelding, *The Diversity of Crop Plants* (London: Academic Press, 1983): 679–700.
2. Linda Perry et al., 'Starch Fossils and the Domestication and Dispersal of Chili Peppers (*Capsicum* ssp. L.) in the Americas', *Science* 315, no. 5814 (February 2007): 986–88, https://doi.org/10.1126/science.1136914; Barbara Pickersgill, 'Domestication of Plants in the Americas: Insights from Mendelian and Molecular Genetics', *Annals of Botany* 100, no. 5 (August 2007): 925–40, https://doi.org/10.1093/aob/mcm193.
3. Carl O. Sauer, *Agricultural Origins and Dispersals*, (The American Geographical Society, 1952): 71.
4. Further details in an entertaining account: Jonathan Sauer, *Historical Geography of Crop Plants* (CRC Press, 1993): 160.
5. Jean Andrews, 'Diffusion of Mesoamerican Food Complex to Southeastern Europe', *Geographical Review* 83, no. 2 (April 1993): 194–204, https://doi org/10.2307/215257.
6. Andrews, 'Diffusion of Mesoamerican Food': 195.
7. Andrews, 'Diffusion of Mesoamerican Food': 194–204.
8. Paul W. Bosland and Jit. B. Baral, 'Unravelling the Species Dilemma in *Capsicum frutescens* and *C. chinense* (Solanaceae): A Multiple Evidence Approach Using Morphology, Molecular Analysis, and Sexual Compatability', *Journal of the American Society of Horticultural Science* 129, no. 6 (2004): 826–32, http://doi.org/10.21273/JASHS.129.6.0826.
9. More about the heat of various chillies and the heat numbers: 'The Scoville Scale', *Alinmentarium*, last accessed 14 March 2022, https://www .alimentarium.org/en/magazine/infographics/scoville-scale.

10. A rather unpleasant video of people doing silly things consuming chillies: 'Competitive Eaters Take on World's Hottest Pepper', *Now This News*, 23 May 2018, https://nowthisnews.com/videos/food /competitive-eaters-take-on-worlds-hottest-pepper.

Not Just for Hallowe'en

1. H.S. Paris, 'A Proposed Subspecific Classification for *Cucurbita pepo*', *Phytologia* 61 (1986): 133–38.
2. U.P Hendrick, *The Vegetables of New York*, vol. 4, (Albany, NY: New York Agricultural Experiment Station, 1928): 3.
3. Hendrick, *The Vegetables of New York*: 47
4. Sturtevant, *Edible Plants of the World*: 245.
5. Thomas W. Whitaker and G.W. Bohn, 'The Taxonomy, Genetics, Production and Use of the Cultivated Species of Cucurbita', *Economic Botany* 4, no. 1 (January 1950): 52–81.
6. Sauer, *Historical Geography of Crop Plants*: 49.
7. Lorenzo Raimundo Parodi and Angel Marzocca, 'Relaciones de la agricultura prehispanica con la agricultura Argentina acutal: Observaciones generales sobre la domesticacion de las plantas, agricultura precolombina y colonial en Latino América: orígenes y promotores' [Relations of pre-Hispanic agriculture with current Argentine agriculture: General observations on the domestication of plants, pre-Columbian and colonial agriculture in Latin America: origins and promoters], *Ann. Nac. Agron.y Vet.Buenos Aries I* (1935): 115–37.
8. Giovanni Ignacio Molina, *Saggio sulla Storia Naturale de Chili*, vol. 1 (Bolgna: Nella Stamperia, 1782): for an English translation, see *The Geographical, Natural and Civil History of Chile* (Cambridge: 1809).
9. 'Heaviest Pumpkin', *Guinness World Records*, last accessed 14 March 2022, https://www.guinnessworldrecords.com/world-records/heaviest-pumpkin.
10. W.F. Giles 'Gourds, Marrows, Pumpkins and Squashes', *Journal of the Royal Horticultural Society* LXVIII, part 5, (1943).

And Finally – Seeds of Hope

1. Delphine Renard and David Tilman, 'National Food Production Stabilized by Crop Diversity', *Nature* 571 (2019): 257–60, https://doi.org/10.1038 /s41586-019-1316-y.

Index

Afghanistan, and carrots, 57, 58–59,
  66–67
Africa
  Centres of Diversity, 15
  chilli peppers, 238–40, 243, 244
  gourds, 257
  peas, 26
  sweet peppers, 249
agriculture
  Aleppo, Syria, 251
  Fertile Crescent, 188
  Hopi tribe, 199
  Mexico, early evidence of, 155, 236
  Peru, and signs of irrigation, 221
  regenerative, 268–70
Algeria, and wild asparagus, 110
All-Russian Research Institute of Plant
  Industry, 14
Al-Nabhani, Nabhan, 141
Anglesey, and wild leeks, 73
Anglo-Saxon Europe, and leeks, 80
'Angry Bean', 217–19, 224
aphrodisiacs
  asparagus, 109
  carrot, 58
  common bean, 190
  garlic, 144, 146
  lettuce, 125

tomato ('love apple'), 169
Apicius, Marcus Gavius, *De re coquinaria*
  (The Art of Cooking), 26, 108
Appellation d'Origine Controlée (AOC),
  230
Arawak Indians, 156, 221–22, 237, 245
archeological evidence
  agriculture in Mexico, 155, 236
  beans in Mesoamerica, 185, 187,
    220–21
  beans in Syria, 42
  chillies and peppers, 235–37
  Guitarrero Cave, Peru, 221
  Huaca Prieta, Peru, 221, 261
  maize, 209
  peas, 25
  Peruvian pottery, 221
  squash, 261
  Sumerian sources, 79, 125–26
  Tell Abu Hureyra, 43
  'Three Sisters' gardens, 186
Asia
  asparagus, 107
  broad beans, origins of, 43
  carrots, 60
  chillies, spread of, 239–40, 243–45
  fava beans, 45–47
  garlic, 147–50

spice trade, 239, 244
asparagus (*Asparagus*)
  breeding, 114–17
  etymology, 107
  health benefits, 109, 120–21
  Roman cuisine, 19, 108
  seeds, 112–13
  species, 109–11
asparagus kale, 91–92
Austen, Jane, 169
Australia, and carrots, 69–70

Ballon, Emigdio, 200
Balsas River Valley, Mexico, and maize, 211
Banga, O., Dutch breeder, 66
Battersea Gardens, 111–12
Battle of Crécy, and leeks, 74, 76, 80
Beadle, George, 'Teosinte and the Origin of Maize', 210–11
beans and common beans
  'Angry Bean', 217–19, 224
  Black Delgado bean, 187–88, 197
  borlotto bean, 182, 192
  cannellini beans, 192
  from Catalonia, 54–55
  Cherokee Trail of Tears, 194
  from Chile, 56
  Columbus and, 221–22
  common beans, 51, 183– 91
  cultivars, new, 52–54
  from Damascus, 44–45
  Fagioli di Lamon (Lamon bean), 181–84, 197
  fava or broad beans, 42–51, 221–22
  French cuisine, 193–94
  Heinz baked beans, 196
  Heritage Seed Library, 53
  historical attitudes toward, 47–48

Hopi Tepary, 205
  Italy, 182, 192–93
  lima, 218, 219–21
  longpod, 52–53
  from Myanmar, 45–47
  navy, 195–96
  New World varieties, 156–57
  pallar, 219–24
  pea bean, 190–91
  in Roman culture, 48
  runner bean, 11, 225–27, 228–30
  sieva (type of lima), 222
  terroir, 230
  varieties, in catalogs, 51–52
  Zolfino, 192
Beard, Mary, 19
Belgium, chicory and endive, 135–36
biodiversity
  asparagus, 119
  beans, in Mesoamerica, 186–88
  decline in past century, 28, 273
  leeks, 83
  native crops, 11, 269–74
  resilience to climate change, 214
  transporting seeds, 272–73
Biodiversity International, 87
biotech and genetically modified plants
  BASF company, and leeks, 82, 83
  Bayer, and leeks, 83
  'Frankenstein food', 179
  intellectual property laws, 82–83
  maize experiments, 211
  tomatoes, 178–79
bird's eye cayenne, 239, 243–48
Black Delgado bean, 187–88
Bock, Jerome, 20
Borlaug, Norman, 7, 8
Bowland, George, 54
Bozzini, Alessandro, 151

Index

brassicas (*Brassica*)
  asparagus kale, 92
  broccoli and cauliflower, 19, 98–101
  Brussels sprouts, 101–4
  cauliflower, in Rome, 19, 100
  etymology of, 86–89
  kale, varieties of, 90–92
  kohlrabi, 96–97
  Mediterranean breeders, 98, 100
  Roman cuisine, 93
  Roman invasion of Britain, 88–89, 107
  seven groups of, 85
Brazil, 190, 238–39
Britain
  apples, 5
  asparagus, 111–14, 121–22
  Battersea Gardens, 111
  beans, common, 51, 52, 191, 196
  brassicas, 88–89
  broad beans, 44, 50–54
  cabbages, 89–90, 94, 96
  cannellini beans, 192
  carrot varieties, 66
  cauliflowers, 100–101
  chicory and endive, 137–38
  Flower and Produce Show, 227
  French Huguenots' gardens, 111
  garlic, 151–52
  kale, 92
  ketchup, origins of, 169
  leek varieties, 81–82
  lettuce, 128–132
  marrow (squash), 252–54
  peas, 26–28, 31, 33
  ramsons, 79, 148
  Renaissance herbals, 27
  Roman invasion, 88–90
  runner beans, 226–27, 228
  salad vegetables, 131–32

  Stenner (runner bean), 228–30
  tomatoes, 161
British Asparagus Festival, 121
British Museum, and ancient Egyptian
  peas, 32
broccoli, 19, 98, 100
Brunfels, Otto, 20
Burns, Robert, 'Halloween', 92
Burpee, W. Atlee, 96

cabbage (*Brassica oleracea*)
  French name for, 87
  January King, 94, 101
  Roman cuisine, 19, 93–94
  sauerkraut, uses of, 94–95
Cadwallon of Gwynedd, 75–76
Caesar, Julius, 108, 122
Caldesi, Giancarlo, 191–92
Campbell-Culver, Maggie, *The Origin of*
  Plants, 76
Canyon de Chelly, 201–2
Cardiff Naturalists' Society, 74
carrot root fly, 68–69
carrots (*Daucus*), 20
  from Afghanistan, 66
  from Australia, 69
  clamping, 67
  colour of, 61–65
  disease resistant, 68–69
  domesticated, history of, 57–58
  Dutch breeders, 63–68
  heritage varieties, 69–70
  House of Orange, 65
  from India, 62–63
  Orange or Sandwich, 66
  origins in ancient times, 58–61
  sugar content, 62–63
  from United States, 68
  varieties of, 65–69

Carters, seed merchant, 53, 69, 70
Carter's Blue Book of Gardening, 96, 84
Catalonia, 54, 37, 176
Cato, on asparagus, 107
Cave of Treasure, and garlic remains, 143
Cennin Pedr (Peter's Leek) (daffodil), *73*
Central Asia, and garlic, 148–49
Centres of Diversity, 14–16
Charlemagne, on carrots, 60
Charles I, 225
Charles V, 182
Charlwood, Geoffrey, 29
Chaucer, *The Canterbury Tales*, 127
chickpeas, 43
chicory (*Cichorium*)
    coffee substitute, 136
    and endives, 134–38
    origins, 135–36
chillies (*Capsicum*), 155–157
    Africa, 238–40, 243, 244
    Asia, 239–40, 243–45
    bird's eye cayenne, 239, 243–48
    Britain, 247
    Carolina Reaper, 247–48
    domesticated, 235–36
    Dorset Naga, 247
    Ginnie peppers, 238
    hottest varieties, 236
    Lemon Drop, 233–34
    Mathania, in Rajasthan, India, 241–43
    Myanmar, 248–49
    origins of, 235–37
    Scoville Heat Units (SHU), 246–47
    Tabasco sauce, 236–37, 245–46
China
    asparagus, 118
    beans, 50, 52
    carrots, 62

garlic, 144–45, 149
Shan maize, 213
Sichuan cuisine, 50
Tien Shan (Heavenly) Mountains, 143, 149
Cichorium (endive) varieties, 136–37
Cock-a-Leekie soup, 81
Colombia, and *cargamanto* bean, 182
colonization of New World, 156–57, 222
Columbus, Christopher, 50, 156–57, 183, 221–22, 237
Consortium for the Protection of the Lamon Bean from the Belluno Valley PGI, 182
Cook, James, 95
corn. *See* maize
Cortés, Hernán, 157, 167–68, 182, 225, 239
Court of Eden gene bank, 34
Culpeper, Nicholas, 242
cultigen, defined, 77
Curry, Ed, and Carolina Reaper chilli, 247–48
Cyprus, and cauliflower seeds, 99

daffodil (*Cennin Pedr*) (Peter's Leek), 73
Daléchamps, Jacques, 127
Damascus, and broad bean, 44–45
Darwin, Charles, 209
Day, Joseph, 206
de L'Obel, Matthias, 222
Devi, Mrs., and Mathania chilli, 241–42
Diogenes, on kale, 87–88
Dioscorides, 19, 72
    De materia medica, 20
diversity. *See* biodiversity
Dodoens, Rembert, Father of Botany, 20, 127, 168
    De frugum historia, 189

Index

Durante, Castor, *Il Tesoro della Sanità* (Treasury of Health), 190

Egypt
  asparagus, 107
  carrots, 59
  common bean, 47, 190
  common leeks, 77–80
  endive, 136
  garlic,143
  lettuce, 125
Elizabeth I, 66, 225
escarole, 138
Ethiopia, and origins of garden pea, 26
EU Common Catalogue of Vegetable Varieties, 67, 83
European General Catalogue of Vegetables, 53
Evelyn, John, *Acetaria: A Discourse of Sallets*, 128
'Experiments on Plant Hybridization'(Mendel), 39*n*
Ezekiel's bread recipe, 47

fava or broad beans
  'Angry Bean', 222, 224
  Damascus variety, 44–45
  favism (poisoning), 49–50
  historical attitudes toward, 47–48
  Palmyra variety, 42–44
  Manipuri cuisine, 50
  Myanmar variety, 45–47, 221–22
  Neolithic culture, 50–51
  Sichuan cuisine, 50
Fertile Crescent, 14
  Arabian Peninsula, 141
  broad beans, 50, 188
  common and wild leeks, 77–78
  lettuce, 124, 125

Flat Holm, island in Bristol Channel, 72
Flower and Produce Show, 227
Folk Varieties (FV), 206–7, 213, 271
F1. *See* hybrid varieties
food security, 196, 270–73
Founder Crops, 42–43
France
  asparagus, 111
  cabbages, 94
  common bean (haricot), 193–94
  escarole, 137
  French beans (haricot), 196–97
  haricot blanc, 227
  leeks, 81–82
  lettuce, 129, 132
  peas, 28, 34
  'peasant fare', 193
  runner beans, 227
  tomatoes, 175
  Vilmorin-Andrieux, 34
Frederick the Great, 136
Fuchs, Leonhart, 20, 97, 189
  New Herbal, 127, 239

Galapagos Island, and wild tomato, 163
Galen, 255
Gardener's Chronicle, The, on lettuce, 131
'Gardens of Adonis, The,' 126
Garraway, Stephen, Seedsman and Net Maker, 51, 96
garlic (*Allium* genus), 145–50
  as aphrodisiac, 144, 146
  Arabian Peninsula growers, 139–42
  Ayurvedic views on, 146–47
  etymology, 145–46
  elephant garlic, 78
  hardneck, 147–50
  medicinal uses, 143–47

garlic (*continued*)
  origins, 143
  performance-enhancing, 144
  propagation of, 148, 151
  protective powers of, 146–47
  scapes, 147
  softneck, 147–50
  studies, 149–50
  wild, 150
genetic diversity
  gene banks, 10, 15
  heterozygosity, 214
  narrow genome, dangers of, 215
genetic modification
  maize cultivation, 215
  regulation of genetic material, 272
George, David Lloyd, 74
Gerard, John, 225
  The Herball or Generall Historie of
  Plantes, 190
Gerhart, Elfrid, 68
Germany
  asparagus, 111, 121–22
  cabbages, 87, 90, 94–95
germination, timing of, 186–87
Gibbon, Robert, 173
Giles, W.F., 265
Ginnie peppers, 238
Gisler, Melanie, 200
Gordon, J., plant catalogue, 66, 171
Great Britain. *See* Britain
Greece
  brassicas, etymology of, 87–88
  carrot and parsnip, wild, 59
  garlic, 143
  lettuce, 12
  runner beans, 227
Green Revolution, 7–8, 9
Grimstone's Egyptian Pea hoax, 32–33

Guilandino, Melchiorre, 167, 168
Gutenberg, Johannes, and printing press,
  20–21

'Halloween' (Burns), 92
Heinz, Henry John, 196–96
heirloom and heritage vegetables
  American vegetables, 265–66
  Britain, 12
  cabbages, 96
  chillies, 245
  Christmas lima bean, 224
  international intellectual property
    laws, 83
  leeks, 75, 84
  open-pollinated, 12, 40, 100
  runner beans, 231
  seed banks, 271
  in Southwest (US), 203, 206, 215
  squash, 254
Henry V, 76
Henry IV, 145
Herbary in Highgate market, 32
Heritage Seed Library, 10, 33, 53
Herodotus
  on garlic, 143–44
  Histories, 19
  on lettuce, 127, 128
Hesychius, names for brassicas, 87
heterozygosity, 214
Hopi Indian agriculture, 198–99, 201,
  204–6
Horace, 137–38
horticulture and the printing press, 21
Horticulture Research International, 82
Horticultural Society of London, 121
Hubbard, Elizabeth, 264
Hubbell, John Lorenzo, 202–3
'Hungry Gap' and cauliflowers, 101

hybrid varieties
  beans, 197
  Brussels sprouts, 103
  cabbages, 96
  carrot varieties, 67–68, 70
  cauliflowers, 100
  chillies, in India, 240–41
  versus cultigens, 77
  natural, and landraces, 119
  pea breeding, 28
  wild peas, 25

Ibn al-'Awwām, 20, 62, 100
Ibn Sayyār al-Warrā, *Kitab al-T.abih˘*
  (Book of Dishes), 61
indeterminate growth pattern, 170
India
  carrot cultivation, 62–63
  celtuce, 126
  chillies, 240–41
  desi (local varieties), 9
  'evergreen revolution,' 9
  garlic, 144, 146–47
  Jaune de Madras (pea), 34–35
  Rajasthan cuisine, 62, 146–47,
    240–43
Institute of Applied Ecology, 200
International Seed Federation, 271
International Treaty on Plant Genetic
  Resources for Food and Agriculture
  (ITP-GRFA), 272
Iran, 60–61, 79
Ireland, 29–32, 87
Irish Seed Savers Association, 29, 32,
  231
Israel, and garlic, 143
Italy
  asparagus, 119
  borlotto bean, 182, 192

cabbages, 94, 97
cannellini beans, 192
cauliflower varieties, 100–101
cucina povera (peasant food), 192
Black Delgado bean, 187–88, 191–92
Denominazione d'Irigine Protetta
  (D.O.P.) status, 175
etymology for brassicas, 87
French beans, 192
kale, 91
Lamon bean, 181–84, 197
lettuce, 130
San Marzano tomato, 175
tomatoes, 170–71, 175
Tuscany, 159–60, 191–99
Zolfino (bean), 192

Jacquin, Nikolaus Joseph von, 240
Japan, 110, 179
Jefferson, Thomas, 33, 128
  Notes on the State of Virginia, 172–73
Jenkins, J.A., and tomato varieties, 166,
  167

kale
  British cuisine, 91
  Greek cuisine, 87–89
  Roman cuisine, 19
Kapuler, Alan, 38–40
Kazakhstan, and garlic, 143
ketchup, 168–69, 195
Kidner, A.W., 115–16
Kidner's Pedigree, 116
Knapczyk, Jack, 152
Knight, Thomas Andrew, 28, 39
kohlrabi, 96–97
Korea, and chicory tea, 136
Kurdistan, and lettuce, 124
Kyrgyzstan, and garlic, 143, 149

Lake Assad, 43
Lamon bean (*Fagioli di Lamon*), 181–83, 197
Landreth, David, Jr, 31, 265
Landreth, Cuthbert, 265
landraces
    beans, 232
    chillies, 242, 248
    maize and FV, 296–97
    natural hybridization and, 119
    teosinte and maize, 211
    terroir of South Wales, 232
Laos, Phosi market, 22–24
Latini, Antonio, 171
leeks. *See also* Welsh leek
    American leek, 79
    biotechnology and, 82, 83
    disease resistance, 82, 83
    domestication of, 78–82
    in *Henry V*, 74
    hybrid varieties, 82
    intellectual property of, 82–83
    Leac-garth (herb garden), 77
    Lyon Prizetaker, 81
    medicinal uses, 73
    open-pollinated varieties, 83–84
    origins of, 76–80
    Roman cuisine, 19, 76, 8
    Scots cuisine, 81
    Usk Show, 71–72
    varieties of, 77–82
    wild, on British islands, 72–73, 76
Leertouwer, Diederick, and asparagus, 112
lentils, 43
lettuce (*Lactuca*)
    in Britain, 129
    butterhead, 129
    celtuce, 126–27
    Chinese cuisine, 126
    descriptions, early, 127–28
    etymology of, 124
    iceberg, 132–33
    Lactuca serriola, 124
    medicinal qualities, 126
    origins, 124–25
    Romaine, 129
    Roman cuisine, 19, 126
    self-pollinating, 133–34
    seven groups, 125
    wild, 128
Liang, Kenneth, 243
Libya, and wild carrot, 68
lima beans, 219–21
Lind, James, *Treatise of the Scurvy*, 95
Linnaeus, Carl, 20, 59, 74, 86, 88, 90
    Cucurbita, 258
    Species Plantarum, 88
    species classification of the tomato, 168
Lovelock, Yann, 102
Lucas, William, seed seller, 65, 95–96
Lydgate, John, 27
Lyon, Ruth, 112

MacNeisch, Richard, 156
Madras curry, 36–37
Maggioni, Lorenzo, 87
maize, 8–9, 15, 155, 188
    hominy, 212
    Hopi Blue, 198, 201, 206
    mound method of growing, 206
    nixtamalization, 212
    origins of, 208–11
    popcorn, 212
    teosinte, wild, and, 209–12
Malaysia, and chilli peppers, 244

Index

Mangelsdorf, Paul, 210
mangetout, 24, 25, 38–40, 39
  in India, 36
  Jaune de Madras, 34–36
  in Renaissance Britain, 27
Manila galleons, and trade with
  Mesoamerica, 223
Manipuri cuisine and fava bean, 50
Margaret Tudor, 64
marrow, 252–54. *See also* squash
Marshall, Charles, *Plain and Easy
  Introduction to Gardening*, 102
Mathania chilli, 241–43
Matthioli, Pietro Andrea, 127, 167
McIlhenny, Edmund, and Tabasco sauce,
  245–46
Meade, James, 169
medicinal qualities of foods
  asparagus, 109
  garlic, 143–47
  Il Tesoro della Sanità (Treasury of
    Health) (Durante), 190
  leeks, 73
  lettuce, 126
medieval agriculture, 20, 59, 94
Mediterranean
  cauliflower and broccoli, 99–100
  kale, domestication of, 90
  leeks, wild, 77–78
Mendel, Gregor, 34, 39*n*
Mesoamerica, 14, 184–88
  beans, 185
  chillies, wild, 235–37
  maize, 211
  squash, 261
  Tabasco and chillies, 245–46
Mesopotamia
  common leek, 77, 79
  lettuce, 125

Yale Babylonian Tablets (cookery
  book), 79–80
Meydenbach, Jacob, 127
Mexico
  archeological sites, 187, 236
  Balsas River Valley, 211
  Black Delgado bean, 187, 197
  chilli peppers, 239
  Fagioli di Lamon bean, 184, 197
  maize, origins of, 209, 211
  New World vegetables, 155–157
  runner beans, 225
  squash, 261
  tomato varieties, 164–65
Michaud, Joy and Michael, 247
Michigan Agricultural Experiment
  Station, 133
Milne, R., 131
Modern Varieties (MV), 213, 241–43
Molina, J., 263
Monticello, 33
Myanmar
  black fava beans (Myanmar), 45–47
  maize cultivars, 213

Native American agriculture
  beans, 194
  chillies, 240
  Hopi people, and blue corn, 198,
    203–7
  maize, 208, 212, 215
  sieva bean, 223
  squash, 255–56, 259, 260–61
  Tesuque seed bank, 200–201
  Three Sisters (corn, bean, and
    squash), 199
Navaho Indians and Canyon de Chelly,
  201–4, 208
navy beans, 196

New, Anthony, 121
'New World' plants, 156–57, 222
New Zealand, and asparagus, 119
Nielsen, Svend Erik, 31
Netherlands
    asparagus, 111
    cabbages, 94–95
    carrot varieties, 65–66
    CGN gene bank, 129
    Dutch breeders, 63–68
    sauerkraut, uses of, 94–95
Northrup King Seed Company, 39
Norton, J.B., 114
nutritive qualities of vegetables, 268

Oman, garlic, 139–42
open-pollinated seeds, 4, 11–12,
    39, 40
    asparagus, 113, 116
    brassicas, 100
    carrots, 67, 68
    leeks, 82, 83, 84
    tomatoes, 166, 177, 178–80
'open source' plant breeding, 40
organic farming
    movement, 9, 161, 269
    sustainability, 267–73
    seeds, 11
Oskam, Gerrit, 34

Pakistan, wheat production, 7
Palmyra, and fava bean, 42–44
Pangalo, K.J., 260
Paris, Harry S., 252–53, 266
peas. *See also* mangetout
    Avi Joan (pea), 37–38
    breeding, 28
    Champion of England, 33
    Daniel O'Rourke, 29–32

De re coquinaria (The Art of
    Cooking), 26
Egypt, 32
Ethiopia, 26
in Fertile Crescent, 43
heirloom, from Laos, 23–25
origins of, 25–26
pigeon, 27
Prince Albert, 31, 33
protein source, 25–26
Sangster No. 1, 30, 31
self-fertile, 27
'peasant food' (*cucina povera*), 183–84,
    187–89, 192, 197
peppers. *See also* chilli peppers
    Aleppo pepper, 251
    Capsicum annuum, 2
    Morocco, 250–51
    pimiento, 249–50
    sweet, 249–51
Pepys, Samuel, on asparagus, 111
Pereira, Diogo Fernandes, 244
Peru
    asparagus, 118
    Fagioli di Lamon bean, 184
    lima beans, 217, 221
    origins of vegetables, 163
    pallar bean, 219–24
    pottery, and motifs from beans, 221
    tomato, wild, 162–63
Picton, Allan, 231
plant breeding, abuses of, 271
Plautus, on brassicas, 87
Pliny the Elder, 19, 43, 47–48, 87, 128
    on asparagus, 107
    on garlic, 144
    Naturalis Historia (Natural History),
    96–97
Poland, and asparagus, 107

Index

Pope Clement VII, 182–83
Portuguese traders
  Africa trade, 243–44
  Asian trade, 243–45
  chilli peppers, 243–44
  Ginnie peppers, 238–39
  imports from New World, 157
  maize traded to China, 213–14
  Manila galleons, 223
potato, from the New World, 157
Protected Geographical Indication
  (PGI), 120, 192
protein, and plants in Centres of
  Diversity, 15
Puckerbutt Pepper Company, 247–48
pumpkin, 265–66
Pythagoras and broad beans, 47, 49

radicchio, 137–38
Rajasthan cuisine, 62, 146–47, 240–43
ramps (US), 79
ramsons (U.K.), 79, 148
Ray, John, 170–71
Read and Hann, seed merchants, 53
Reeves, Robert, 210
Regal, Edward August von, 148–49
Rick, Charles, 164
Rodrigues, Diogo, 244
Roman Empire
  asparagus, 108
  brassicas, 87–89
  broad beans, 48, 50
  carrots, 59, 60
  De re coquinaria (The Art of
    Cooking), 26, 108
  food culture, 80
  garlic, 144
  leeks, 80
  lettuce, 126–27

radishes, 126
Romania, leek and cultural identity, 81
Royal Horticultural Society, 33, 174
Ruel, Jean, *De natura stirpium* (The
  Nature of Plants), 94
Russia
  asparagus, 107
  celtuce, 126
  garlic, pink, 152
  kale, 91

St David, 75–77, 80
Saiq Plateau, in Oman, 140–42
salads, 131–32, 134
Sanders, T.W., *Vegetables and Their
  Cultivation*, 48
Sangster, Joseph, 30, 31
San Marzano tomato, 175
Sauer, Jonathan D., 219, 221, 261, 266
sauerkraut, uses of, 94–95
Schöffer, Peter, 127
Scotland
  asparagus kale, 92
  chicory, 136
  kale soup, 92
  leeks, 81
Scoville, William, 246
Scoville Heat Units (SHU), 246–47
Sichuan cuisine and fava beans, 50
seed libraries, 15
  Folk Varieties (FV), 271
  Leningrad, 14
  protection of heirloom varieties, 271
  Pueblo of Tesuque seed bank,
    200–201, 208
Seed Savers Exchange, 83
seed saving
  diversity and resilience of seeds,
    214–15

seed saving (*continued*)
early agriculture, 4–5, 11, 14–16
Hopi culture and maize, 206–8
selection, in early agriculture, 5
swaps, 15–16
Severn Hills, 103
Shan State, Myanmar, 213, 216–19, 224
Shekhawat, P.S., 240–41
Sharpe, John, 199–200
Simon, Philipp, 149
Slow Food USA's Ark of Taste, 219
Smith, Brian, 82
Soe, Ah, 217–18
South America. *See also* Peru
Argentina, and squash, 262
Bolivian cuisine, 235, 263
Chile, and lima bean, 220
chillies, 237
origin of plants, 156–58
South Wales, *terroir* (for beans), 231–32
Southwest (US)
agriculture, 199–204
chillies, 240
Hopi Blue maize, 198, 201, 206
squash, 261
Spain, 61–62
asparagus, 110, 119, 120
Avi Joan (pea), 37–38
Catalonian organic farm, 37
chillies, from the Bahamas, 236
radicchio, 137
tomatoes, 175–76
Spargelfests (asparagus festival), 121–22
spice trade, black pepper and chillies,
239
squash (*Cucurbita*). *See also* marrow
bottle gourd, 255–57
etymology, 255
four types of, 255–57

Heritage Seed Library, 264
hybrid, 262
Native American cultivation, 255,
259
pumpkin, defined, 265–66
seeds, uses of, 256
Städler, Thomas, 165
Stenner, Brython, 228–30
Sturtevant, Edward Lewis
Sturtevant's Edible Plants of the Wild,
92, 94
Sudan, and asparagus, 110
Sutton & Sons, plant breeders, 107
Swaminathan, Mankombu, 9
sweet potato (*Ipomoea*), 156
Switzerland, and cabbage, 94
Syme, James, 89–90
Syria
Aleppo pepper, 251
fava or broad bean, 42–44

Tabasco sauce, 236–37, 245–46
Tabqa Dam, Syria, 43
Tamaro, Domenico, *Orticultura*, 253
Taree or Persian Leek, 78–80
taxonomy of plants, 20
Tell Abu Hureyra, and early cultivation,
43
teosinte and maize, 209–12
terroir (for beans), 231
Theodorus, Jacobus, 127
'Three Sisters' (squash, beans, and
maize), 188–89, 200
Tian, Feng, 208
Tilman, Bill, 181
tobacco, 156, 246
tomatoes (*Solanum*)
as aphrodisiac, 169
Aztec names for, 166

etymology, 165, 167–68
flowers, described, 171
heritage and heirloom varieties,
    160–61, 177
hybrid varieties, 166
genetically modified, 179
indeterminate growth pattern, 170
Italian cuisine, 170–71
ketchup, 168–69, 195
'love apple', 169
Mayan cultivation of, 154
from the New World, 157, 162–63
open-pollinated versus hybrid,
    178–80
plum, in Tuscany, 159–60
pollination, described, 171–72
toxicity of, 161–62
Trophy (hybrid), 173–74
United States, 172–74, 177
Tomato Genetic Resources Centre, 164
Tomato Trust, 177
Tradescant the Younger, John, 226
Turkey, 61, 150
Tuscany, 159–60, 191–99

United States. *See also* Native American
    agriculture
    asparagus, 112, 116, 118
    baked beans, 194–95
    cabbages, 93, 96
    common beans, 190, 194
    Heinz baked beans, 196
    Hopi Blue maize, 198, 201, 205–6
    kale and collards, 91
    leeks, 83
    lima beans, 223–24
    navy beans, 195–96
    organic movement, 9
    peas, 28, 30, 31, 35–40

Southwest agriculture, 199–204, 240,
    261
    squash, 260–64
    tomatoes, 172, 176
    Tomato Trust, 177

Underhill, Ruth Murray, *Workaday Life
    of the Pueblos*, 207–8
US Department of Agriculture (USDA)
    asparagus, 113–14
    leeks, 81
    lettuce, 130, 132–34
    1937 *Yearbook of Agriculture*, 33–34
    pea varieties, 28
    plant breeding in 19th century,
        132–33

Vachell, Eleanor, *The Leek: The National
    Emblem of Wales*, 74–75
Valeriano, Piero, 182–83
van der Donck, Adriaen, *Description of
    the New Netherlands*, 112
Vargas, Jesus, 37–38, 54, 176
Vavilov, Nikolai, 14–15, 29, 30, 60–61
Vilmorin-Andrieux, 34–36, 96, 132
    endives (*Frisée*), 137
    tomato varieties, 174

Van Eseltine, G.P., *Vegetables of New
    York*, 256
veg-box projects, 269

Wade, B.L., 34
Wafer, Annie, 219
Waite, a Massachusetts pea breeder, 29
Wales, 71–77, 80. *See also* South Wales
Waring, George E., 174
Watt, George, 239
Weaver, William Woys, 33

Webb, John, seed merchant, 171
Welsh leek, 72–75, 80. *See also* leeks
wheat production
    Carter's Blue Book of Gardening, 84
    einkorn and emmer, 43
    Green Revolution, 7–9
Whitaker, Thomas W., 133
Willemijns, Mathias, 265
Williams, Llewelyn, 75
Winch, Tony, *Growing Food*, 219

Winthrop, John, Jr., 95
World Fair (Chicago), 30, 31

Yale Babylonian Tablets (cookery book),
    79–80

Zea, and origins of maize, 209–10
Zingiberaceae, and Ginnie peppers, 238
zucca (marrow), from Italy, 253–54
zucchini (courgette), 253–54